Ma

The Quays and Bridges of Paris

An Historical Guide

MARTELLE
É.D.I.T.I.O.N.S

Cours la Reine
Q. des Tuileries
LA SEINE
Cours Albert 1er
Q. d'Orsay
Pt. de l'Alma
Pt. des Invalides
Pt. Alexandre III
Pt. de la Concorde
Pass. de Solférino
Q. Anatole-France
Pt. Royal
Trocadéro
Pass. Debilly
Pt. d'Iéna
Q. Branly
Hôtel des Invalides
Ave de New-York
Pt. de Bir-Hakeim
Ecole Militaire
Radio France
Q. de Grenelle
Ave.Pt. Kennedy
LA SEINE
Pt. de Grenelle
Pt. Mirabeau
Q. Louis Blériot
Q. A. Citroën
Pt. du Garigliano
LA SEINE
Q. de l'Horloge
Q. de Corse
Palais de Justice
Q. des Orfèvres
Q. Marché-Neuf
Parvis N.Dame
Quais de l'Ile de la Cité et de l'Ile Saint-Louis

LES PONTS & LES QUAIS DE PARIS

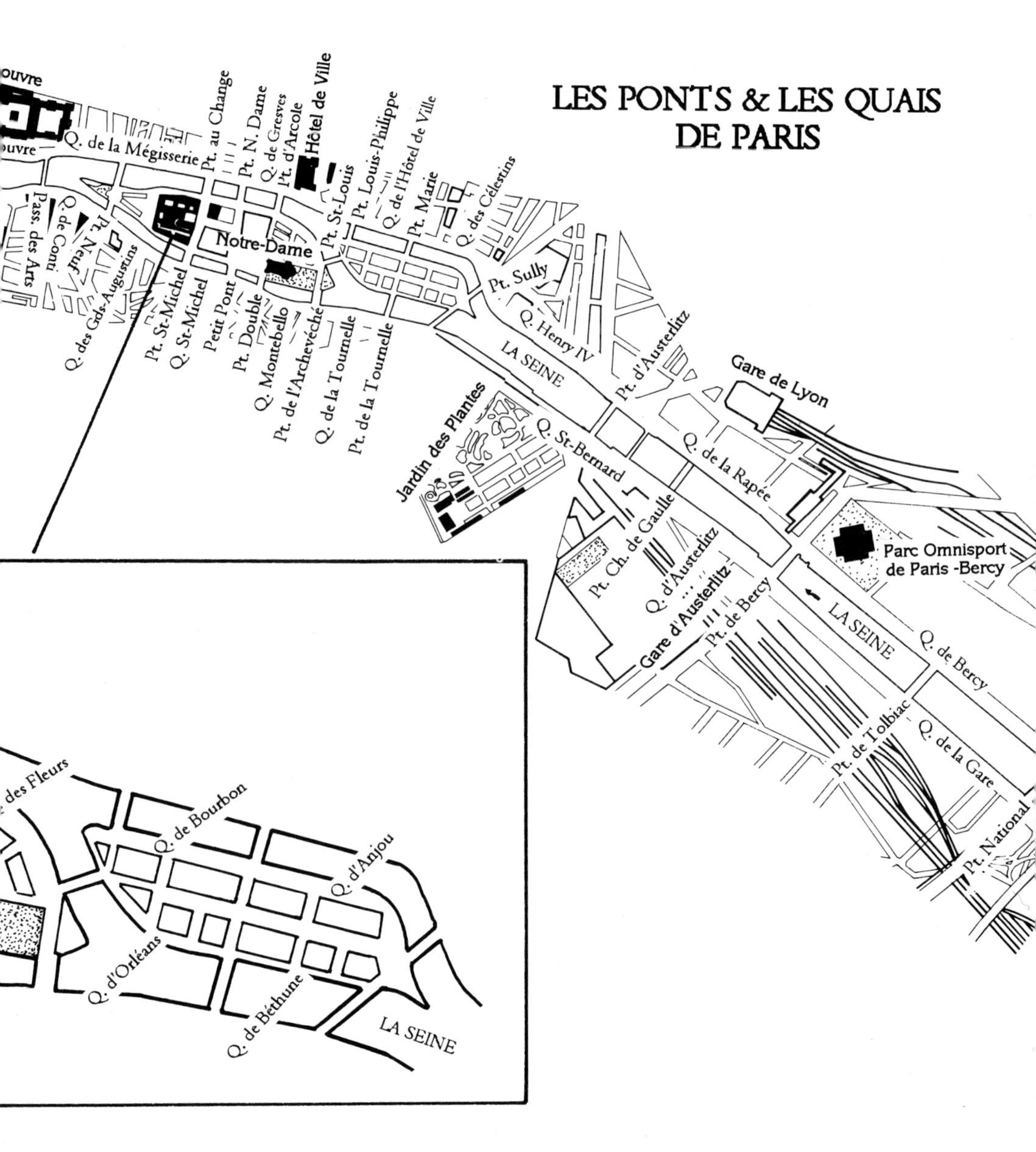

CONTENTS

« In none of the towns that I know do the quays resemble those of the Seine in Paris, except perhaps the quays of the Arno in Pisa. Montaigne noticed this long before me. I had not yet read the Voyage en Italie, by the author of the Essais, when I noticed this similitude. It was perhaps even more noticeable then than today, and the vigilant man strolling through the Augustins and the Lungarno Region perceived it more readily. Because at that time, the riverside landscape of Paris was as mineral as the Italian; private or monumental architectural structures, at least where accessible to the public, were not accompanied by greenery, by plantations; Nature's artifices were not added to complement these works of art; leaves did not readily break the hard lines of stonework. If, at the time of Concini and Mazarin, the Peninsular Invasions brought fragments of Florence to the Luxembourg Palace, Romanized and «Mediterraneanized» certain aspects of our city, so, later on, on the other hand, English influence and Napoleon III, a great lover of gardens, and a highly competent one too, did not spare the sentimental and botanical benefits of Sylvan Romanticism. In certain parts Tuscany and the pontifical city join and mingle with Hyde Park, in particular on the Quay des Augustins. The entablature of the Hôtel des Monnaies (the Mint), a dwelling associated with Michelangelo and the Palais Farnèse, dominates the rows of fragile aspen, melancholy, trembling and shimmering in the breeze while a sun, too weak to dry out the surfaces of the buildings and to bathe them in golden light, plays lovingly with the waters and the leaves - emotive and spasmodic. The Roman Palace and the aspen, Hercules and Ophelia; these combinations seem natural. And, between the stonework and the boughs, separating them with a greenish border, the stalls of the bouquinistes (booksellers) on the parapets, the winding, long, thin rampart of printed pages, of yellowed ink, of damaged bindings, the curb of scholarly kelp, literary debris left on the shore by the ebbing tide of a civilisation where writing reigned, where neither speech nor image propagated, where the subterfuge of the alphabet imposed itself upon all means of communication.»

Alexandre Arnoux.
Paris ma grande ville.

Introduction

To tell the story of the quays and the bridges of Paris within the limits of a single book is an impossible wager, and all the more so of a guide-book. That would mean summarising the history of Paris during the 2,000 years that it has been expanding up from the Seine.

This book therefore proposes an historic panorama and a route through the river town of today, from the banks of the Seine to those of the Canal Saint-Martin; a route which we hope will make the reader's discoveries easier.

There have been so many transformations of the banks of the Seine that we have thought it interesting to conjure up some of their appearances through the iconography of the city, since the text itself refers to these historic aspects of a near or distant past.

Long before UNESCO included the Seine and the quays of Paris in its list of the world's great sites, everybody knew that they were, here, in the heart of one of the most beautiful sites in the world; a site which this guide invites you to discover or re-discover.

The Seine and the Pont National seen from the banks of Ivry, circa 1880.
Painting by Stanislas Lepine.

The pont National

For a century this work marked the upstream limit of the city of Paris, that is, up to the construction in 1967-1968 of a viaduct for the ring road.

Built in 1852-1853 the work was called the Pont Napoleon III. At 240 metres long, it is the longest bridge in Paris. It is 34 metres wide. It is made up of 7 masonry arches with 34-metre spans, two of which are above the quays.

This bridge starts on the right bank at the end of the Boulevard Poniatowski, above the Quai de Bercy, and ends on the left bank, on the Boulevard Masséna above the Quai de la Gare. On the Paris side the work carries the peripheral railway track, separated from the road bridge by railings.

On the right bank there is still one of the ancient «redoutes» (strongholds) of the Fortifications which stretched 40 kilometres round Paris, varying in width from 6 to 700 metres.

This wall, the biggest fortification in the world after the Great Wall of China, was built between 1840 and 1845 to protect Paris.

View of the Pont Tolbiac and the Pont National, taken from downstream to upstream.

The pont de Tolbiac

190 metres long and 20 metres wide, this bridge opens on the left bank onto the street of the same name and on the right bank onto the Rue de Dijon. It commemorates Clovis' victory of Tolbiac in 496. Its construction began in 1879 on the occasion of the construction of the new Quai de Bercy. It was finished in 1882.

It is a beautiful masonry work with five elliptical arches. The central arch spans 35 metres, the next two span 32 metres and the outer two span 29 metres.

The 4-metre thick piers are protected on each side by semicircular masonry pier heads. The embossed masonry continues on the end arches and on the wide pilasters which separate the arches. The parapets are made up of ashlar balustrades.

Tugs on the Quai de Bercy, circa 1890.
Water-colour by Fraipont.

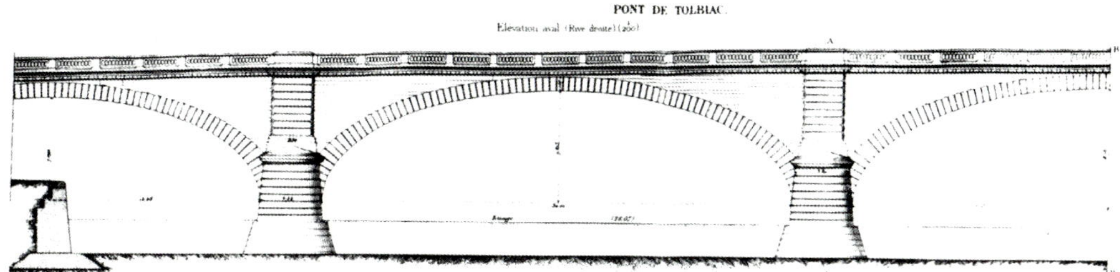

The quai de Bercy

Unloading of goods on the quai de Bercy, circa 1890. Water-colour by Fraipont.

Begins at the Pont National
and ends at the Pont de Bercy

This quay has existed since the end of the XVIIth century. At that time the construction of country houses was beginning at Bercy and later, a large castle was built for the Malon family by the architects Mansart and Louis Le Vau. André Le Nôtre landscaped the gardens. The quay disappeared in the middle of the XIXth century with the setting up of a big wine warehouse and the construction of the PLM railway (Paris Lyon Méditerranée).

Bercy castle and its park, in the XVIIIth century. An engraving by Carnavalet.

A street among the warehouses of Bercy, circa 1890. A drawing by Fraipont.

From 1860 onwards, this warehouse occupied the entire perimeter delimited by the quay, the Boulevard and the Rue de Bercy. It was surrounded by railings. For 130 years until its closing down in the 1970s-80s, all the great wines of France could be found there.

Near the Pont de Bercy, the new Palais des Sports of Paris was built in 1980-82 by the architects Michel Andrault and Pierre Parat. Nearby a beautiful fountain, in the form of a cascade, was built by the sculptor Gérard Singer in 1980.

A park has replaced the former storehouse and new housing developments have been built along the edge of the park (and Rue de Bercy), while a «Cité du Vin» is being built on the upstream part: a warehouse where a certain number of wine and spirits stores have been kept to bear witness to the past. During these works, in August 1991, an important discovery was made; three oak dug-out canoes, a bow and about fifteen hundred miscellaneous objects such as weapons and tools dating back to the so-called «chassean» period (between 4,200 and 3,400 BC).

In the past, the Quai de Bercy was pleasantly enlivened by open-air cafés with music and dancing where the Parisians used to go on Sundays and other non-festive days to savour the vintage (and other) wines... With the current development, new riverside restaurants and open-air cafés will be set up. The quay and the park will be linked to the Grande Bibliothèque by a foot-bridge; it will cross the Seine upstream of the Pont de Bercy.

In the line of the Rue de Macon with its two rows of plane trees, opposite the Pont de Tolbiac, is the Notre Dame de Bercy church built by the architect Molinos in 1824.

Along the edge of the park, on the Rue de Bercy side, the architect Franck O. Géhry built, in 1993, the American Cultural Centre of Paris.

On the bank, the new facilities of the Port de Bercy were completed in 1991-92 by the architect Luc Arsène-Henry: premises for showrooms and the sale of building materials, the «Bétons de Paris» power station, the silo of the «Sablières de la Seine»...

Dug-out canoes of the Chassean period (4200-3400 B.C), discovered in Bercy in 1991.

The Coopers of the Rue Laroche in Bercy, circa 1880. Water-colour by Fraipont.

The multi-purpose sports centre at Bercy. Andrault and Parat, architects, 1981.

The quai de la Gare

Begins at the Pont National
and ends at the Pont de Bercy

*The station gate, circa 1890.
Drawing by Fraipont.*

In the XVIIth century this quay was a towpath. It later became the entrance into Paris of the main road RN 19 (Paris-Bâle) and therefore had to be paved. Its name comes from the presence of a dock or «gare d'eau» opened under the reign of Louis XVI in 1764, to receive boats transporting cereals and flour. Later, the «Grands Moulins de Paris» were also located here. At the beginning of the XIXth century, the Saget glassworks, established before the Revolution, could still be found there.

The iron bridge which spans the railway tracks perpendicular to the Pont de Tolbiac is an unusual construction made of bolted steel girders. This work, erected in 1895, belongs to the urban heritage of Paris. It was designed by the engineer Henri Daydé who was also involved in the construction of the Pont de Passy and the Pont Mirabeau. On the quay, not far from the viaduct, are the ancient buildings of the Grands Moulins de Paris, notably the grain and flour silos whose reconversion seems difficult: certain elements could perhaps be retained or transformed into a souvenir of the past.

At the beginning compressed air was used for lifts and to adjust the hands of public clocks: one impulse every second.

There are no outstanding ancient buildings on this quay. We should however point out, not far from the Pont National, the factory of the «Société Urbaine d'Air Comprimé» (Urban Compressed Air Company), built in 1890 by the engineer Joseph Leclaire; a factory whose large hall is a remarkable iron construction. It is evidence of the industrial vocation of this district from the beginning of the XIXth century to recent times.
Not far behind the quay, passing under the railway tracks, the Rue Watt in one of the strangest streets in Paris with a somewhat dreary and mysterious atmosphere, often chosen as the setting for the shooting of films adapted from thrillers ...

*The factory of the Société Urbaine
de distribution d'Air Comprimé,
façade detail.*

A huge new district is soon to replace the warehouses and buildings on the Quai de la Gare. The railway tracks will be covered by a garden

built onto a concrete base; an avenue will run from the Gare d'Austerlitz to the very edge of Paris, in the direction of Ivry.

The Quai de la Gare and the Grande Bibliothèque de France (model). Architect Dominique Pérault.

The construction of the Très Grande Bibliothèque de France – the TGB – annexe of the National library, was started in 1992. Based on the designs of the architect Dominique Pérault, this enormous building fills the area between the Pont de Tolbiac and the Pont de Bercy. Several million books will be at the disposal of the public and research workers, who will benfit from the computerisation of the catalogues and book-borrowing systems.

The Bibliothèque de France will be linked to all the libraries and cultural facilities throughout the country as well as to all the leading universities in the world. The building which has raised polemics will nevertheless be a prototype, in the vanguard of modernity, in both functional and architectural aspects.

The Bassin d'Austerlitz. Painting by Guillaumin.

The pont de Bercy

The Pont de Bercy and the Metropolitan Viaduct, circa 1920.

At right angles with the Boulevard de Bercy on the right bank and the Boulevard de la Gare on the left bank, this 175-metre long, 35-metre wide bridge bears a magnificent stone viaduct built for the «Etoile Nation» Metro line in 1906. This viaduct has 35 semicircular stone arches in ashlar.

The Pont de Bercy was built extremely quickly, in little more than a year, in 1863-1864, under the supervision of the engineers Romany, Savarin, Meunier and Warest. It replaced a toll suspension bridge.

The Pont de Bercy has five elliptical arches with a span of 29 metres;

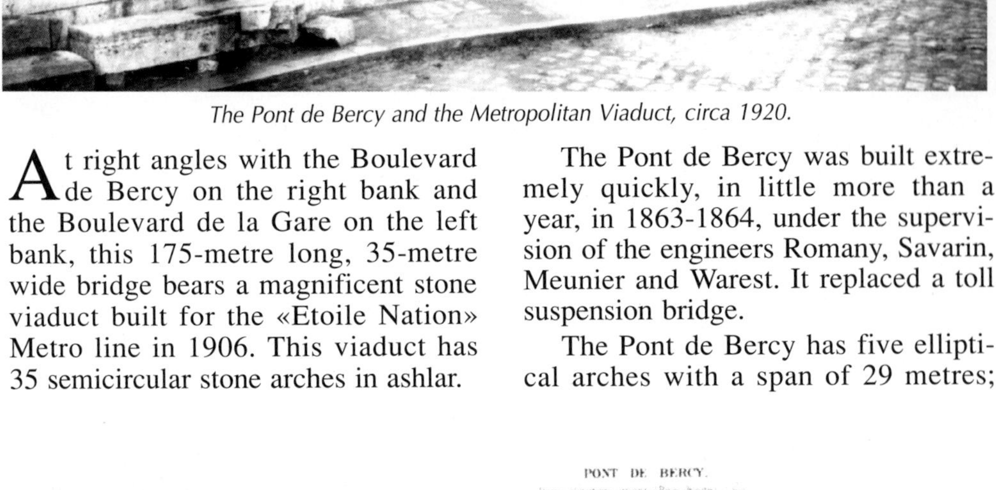

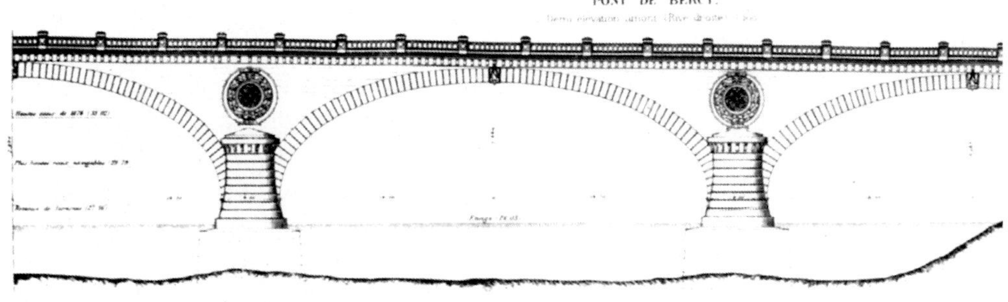

The Pont de Bercy and the Treasury.

the abutments measure 7.5 metres and the piers are 4-metres thick. The piers are protected by pier heads with semi-circular bosses, and have a very graceful profile. Above, laurel wreaths are sculpted in high relief around a central floral motif. At the keystone of the outer arches there is the monogram of Napoleon III. On the cornices the guard-rails are of moulded cast iron.

Until 1990, it was 19 metres wide, but with the building, in 1993, of a new roadway on the upstream side, it was extended to 35 metres.

A figurehead above the bank of the Quai de la Rapée, the Treasury.

The pont Charles de Gaulle

The construction of this new bridge started during the Summer of 1993, perpendicular to the railway stations Gare de Lyon and Gare d'Austerlitz.

It will be a perfectly horizontal continuous superstructure with three spans of 68, 84 and 55 metres resting on two piers in the riverbed.

The 34.9-metre wide superstructure will be formed by a central supporting box section, and two lateral wings; a prestressed concrete slab will form an 18.2-metre wide roadway (6 lanes). There will also be two 5-metre wide footways and two 1.7-metre wide refuge footways higher than the two former ones, and two 1.65-metre balustrades.

The superstructure will rest on the piers on four large, hollow, truncated capitals made of moulded steel.

The piers anchored in the chalk riverbed, will be faced with polished white architectonic concrete. The abutments will be made of reinforced concrete. Steps on either side of the bridge will enable pedestrians to pass between the upper and lower quays. The balustrade facings on the stairways and the adjacent quay walls will be in stone.

The profile of the bridge has been calculated to give an impression of finesse above the water. The work, with its slightly convex lower part, will be entirely covered with wide metal sheets, with very thin joints, and protected by an off-white matt paint.

Its distinctive design will enhance the corolla structure of the bearing capitals and will emphasise the perfect horizontality of the two balustrades bordering the wings of the superstructure.

The station and the Quai d'Austerlitz, circa 1870.

Model of the Pont Charles de Gaulle.

The design of the Pont Charles de Gaulle was the subject of a competition in 1989. The winning project is the work of the architects Louis Arretche and Roman Karasinski. The «Agence des Grandes Opérations and the Direction de la Voirie of the Mairie de Paris» (the Department of Civil Engineering) is supervising the work.

This bridge is in the line of the Rue Van Gogh on the right bank; later it will be extended on the left bank by a viaduct crossing over the Gare d'Austerlitz and bordering the Salpêtrière Hospital in the direction of the Boulevard de l'Hôpital and the Boulevard Saint Marcel. Its creation will necessitate a complete reorganisation of the roadways on the Quai de la Rapée.

The banks, at present occupied by harbour facilities and building material depots, will become a promenade, whereas the quay itself will be entirely replanted. Two subways will allow access to it from the buildings of the Quai de la Rapée.

The quai de la Rapée

Begins at the Pont de Bercy
and ends at the Pont Morland.

This quay owes its name to the presence, as early as the XVIth century, of a country house, called «Hôtel de la Rapée». Jean-Baptiste La Rapée was an Army General Commissioner who served under Louis XIV. For several centuries up to the 1900s, this quay was one of the most important wood ports of Paris for planks, timber and firewood.

On the upstream side, near the Pont de Bercy, the new Treasury, inaugurated in 1989 is the work of the architects Chémètov and Huidobro. This long building is symbolically situated on the site of the ancient toll gate of the «Fermiers Généraux» (Farmer Generals). It is the sole building in Paris to be built directly over the river; it extends over the quay, a symbolic gesture, recalling the age-old symbiosis of the town and the Seine...

The Quai de la Rapée
by Michel Heude de l'Hay.

Nearby, there is a harmonious neo-classical building. Semicircular arcades open out on the ground floor. The two storeys of equal height are crowned by a prominent cornice. It used to house the administrative services of the Port de la Rapée, and later, in 1920, became the Ministry of Ex-Servicemen.

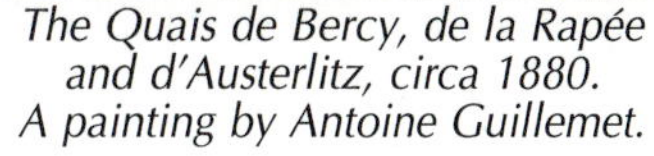

The Quais de Bercy, de la Rapée and d'Austerlitz, circa 1880. A painting by Antoine Guillemet.

At present, the quay is under full transformation. On the site of the ancient Métropolitain electric power station, built at the beginning of the

The Quai de la Rapée and the Pont d'Austerlitz in 1910. A water-colour by René Leverd.

The Quai de la Rapée in the snow. A painting by André Sinet.

The Quai de la Rapée and the Métropolitain power station, circa 1910.

century by the Baron d'Empain, will be, as early as 1995, the Maison des Transports; in particular there will be the offices of the General Management of the RATP, and the Urban Transport Museum , with its ancient vehicles, its models and its archives.

At the end of the Rue Van Gogh, is the high square bell tower of the Gare de Lyon, finished in 1900 during the reconstruction of the station. The faces of this belfry bear the four huge dials of the SNCF's clocks (French Railways).

The Quai de la Rapée and the Clock Tower of the Gare de Lyon.

The flooded Quai de la Rapée in 1910.
Water-colour by René Leverd.

At right angles to the Pont d'Austerlitz, the quay is called Place Mazas - between the Boulevard Diderot and the Boulevard de la Bastille. Jacques François Mazas (1765-1805) nicknamed «Le Brave» was a colonel in the Imperial Army. He was killed at the battle of Austerlitz. After the building of the first bridge, the square was named after this colonel who also distinguished himself by fighting beside La Fayette during the American War of Independence.

Between the Boulevard Diderot and the Avenue Ledru-Rollin, a building has recently been erected in a resolutely contemporary spirit, by the architect Aymeric Zubléna. An elegant steel and glass façade curves behind a huge 85-tonne sliding gate which takes up the whole height of the building. This façade provides an elegant contrast with the neighbouring buildings, constructed in the 1900s.

The quai d'Austerlitz

Begins at the Pont de Bercy
and ends at the Pont d'Austerlitz.

The Quai d'Austerlitz, circa 1880.

This quay was built and paved in the XVIIIth century. Until 1806, it was called the «Quai de l'Hôpital» because of the proximity of the Hôpital de la Salpêtrière.

On this site, the architect Collet built the landing stage of the Chemin de Fer d'Orléans. Inaugurated in 1843, it was replaced in 1867 by the Gare d'Austerlitz, built by the architect Renault.

In the past, the river Bièvre joined the Seine slightly upstream of the Pont d'Austerlitz. It was rechannelled during the construction of the railway landing stage. It has become one of the biggest water courses in Paris, collecting the drain waters of the left bank which are then siphoned across the Seine at the Pont de la Concorde and the Pont d'Alma and directed to the treatment plants at Clichy, and Achères further on.

The Hôpital de la Pitié Salpêtrière, in the background of the Quai d'Austerlitz, was built by the architects Louis Le Vau, Le Muet and Liberal-Brulant at the request of Louis XIV. This hospice was, at the time, an example of modernity with its many well-aired courtyards.

The Quai de la Gare and the Hôpital de la Salpêtrière, as seen from the Quai de la Rapée in the mid XVIIIth century. A painting by Martin Le Jeune. Musée Carnavalet.

The bank of the Quai d'Austerlitz.

The chapel is a perfect example of XVIIth century architecture. Today it is essentially used for exhibitions and concerts. Unique in Paris, it is in the form of a Greek cross crowned by an octagonal dome, itself topped with a lantern.

The Quai d'Austerlitz has been cluttered for half a century by long warehouses built long ago for the Office Customs and Excise; these ungraceful buildings, which block the view of the station and the Hôpital de la Pitié Salpêtrière, are destined to disappear shortly. Their demolition has started at right angles to the Pont Charles de Gaulle. A complete reorganisation of this quay and its banks is being considered.

Upstream of the Gare d'Austerlitz there are the installations of a pumping station built in 1863, on the site of a former «fire pump». Together with those of the Quai Louis Blériot and the Place Mazas, it pumps back non-drinking water for the cleaning of the streets and the watering of parks and gardens.

The Gare d'Austerlitz.

The viaduc d'Austerlitz

The Métro crossing the Viaduc d'Austerlitz. (RATP document).

The Viaduc d'Austerlitz, opened in 1904 between the Quai de la Rapée and the Gare d'Austerlitz, for the passage of the Métro, is one of the most audacieus iron works of art built in Paris. It was designed by the engineer Louis Biette with the architect Jean-Camille Formigé.

Until the construction of the new Pont Charles de Gaulle, it held the record of the longest span; a 140-metre clear span built diagonally across the river. The superstructure which carries the Métro railway tracks, is suspended from two big, triple-jointed, metal arches. This graceful structure includes much moulded iron decoration at the springs of the arches, on the balustrades and at the springs of the piles bearing the superstructure.

The Viaduc and the Pont d'Austerlitz.

The pont d'Austerlitz

A bridge intended to link the Jardin du Roi (The Jardin des Plantes), the Botanical Gardens, to the Faubourg Saint Antoine was designed just before the Revolution by Jean-Rodolphe Perronnet, the architect of, amongst others, the Pont de la Concorde and the first Pont de Neuilly. It was never built.

In 1801, construction was undertaken following the designs of the engineer Lamandé. This bridge, ended in 1805, after the Victory of Austerlitz, was named after the small village in Moravia (Czechoslovakia) where Napoleon's armies conquered the Russians and Austrians.

It was a toll bridge, with five 32-metre iron arches resting on masonry piers. At only 13 metres wide, it was very congested. In 1854 it was extended to 18 metres; the iron arches then being replaced in masonry.

In 1884, the Pont d'Austerlitz was widened a second time to 30.6 metres. All the old stonework was kept. The bridge now has five 32-metre elliptical arches. Above each pier, protected by pier heads with semicircular bosses, large allegorical high relieves, flags, ivy branches and fasces are decorated with masks in cartouches crowned by lions' heads. They are the works of the sculptor Hamel. A solid stone, ashlar parapet runs above.

The Quai de la Rapée and the Pont d'Austerlitz, circa 1883.
A water-colour by Harpignies.

The quai Saint-Bernard

Begins at the Pont d'Austerlitz and ends at the Pont Sully.

The garden of the Quai Saint-Bernard. A marble sculpture by Cardénas.

This old tow-path was first called «Quai hors Tournelles», being situated upstream of the ancient Porte de la Tournelle. It was then given the name of Quai Saint-Bernard because of the presence nearby of the Couvent des Bernardins (Convent of the Bernardines), whose Gothic refectory still exists, converted into a fire station, in the Rue de Poissy. The Quai Saint-Bernard was widened under the Empire, and again in 1839.

The Jardin des Plantes (the Botanical Gardens) occupies the upstream part between the Place Valhubert and the Rue Cuvier. It was created by Louis XIII and his physicians, Herouard and Guy de La Brosse, on a site donated by the Abbaye Saint Victor. The first of its sort in the world, the Jardin des Plantes had a scientific and educational purpose. It was opened to the public in 1640. Jussieu, Lacépède, Daubanton, Geoffroy Saint-Hilaire and Buffon were its most famous Intendants. Buffon directed the establishment for half a century, from 1739 to 1788, and doubled its surface area.

It is from the Place Valhubert - named after one of Napoleon's generals, killed at the battle of Austerlitz - that we have the best overall view of the Garden and its surrounding buildings. There are extensive collections: 15,000 botanical plants; 100,000 species of animals and insects in the long zoological gallery which has been under reorganisation since 1990; two million exhibits in the mineralogy gallery; a million volumes in the library ...

The menagerie of the Jardin des Plantes, near the Quai Saint-Bernard, at the corner of the Rue Cuvier, was created by decree in 1793. The first giraffe seen in France, offered, in 1826, to Charles X by Mehemet Ali, was kept there. It is managed by the Muséum d'Histoire Naturelle, a Public Establishment devoted, since its origin, to scientific research. The eastern part of the Jardin des Plantes, between the quay and the main palaeontological and zoological galleries, is a «jardin français», arranged in a French style and, in the

The Jardin des Plantes and the statue of Lamarck, circa 1830...
...and in 1993.

The Wine Store, circa 1880.
A drawing by Fraipont.

western part, towards the rue Cuvier, is a «jardin anglais», a landscape garden. A great number of statues, most of them dating from the end of the XIXth century, and sculpted by excellent artists, are placed along the paths, the flower beds and in the façades of the buildings.

Among the Museum's buildings, the mineralogical gallery, towards the Rue Buffon and the palaeontological gallery (by the architect Dutert) are evidence of the audacious iron architecture of the XIXth century.

The perimeter between the Rue Cuvier and the Pont Sully, right up to the rue des Fossés Saint Victor, was occupied from the XIth century to 1790, by the Abbaye Saint-Victor, one of the largest abbeys in Paris, destroyed during the Revolution.

Napoleon had a wine store built on its site, majestically rebuilt under the Second Empire (between 1664 and 1789, the wines for the stallholders were already stored on part of this site).

A few rare remains of these imposing vaulted cellars were retained during the building of the «Faculté de Lettres de Jussieu» in the sixties. In one of these beautiful galleries a mineralogy museum has been set up, exhibiting a selection of exceptional crystals and minerals collected from the entire world.

Throughout the XIXth century, and up to the 1930s, the banks of the Quai Saint-Bernard were busy with the work of coopers and the temporary storing of barrels. The landing stages of the «Entreprise Générale des Coches d'Eau», a passenger barge company which used to transport passengers by canal to the centre of France and the Rhône, were also to be found there. At the end of the quay, the Institut du Monde Arabe (the Arab World Institute) was built in 1985, an illustration of functionalist architecture, originating also from avant-garde building techniques.

«Cybernetic» sculpture
by Nicholas Schoeffer.
Jardin Saint-Bernard, 1975.

The Mineralogical Gallery of the Museum is open every day, except public holidays, from 10am to 5pm, and on Saturdays and Sundays from 11am to 6pm (Tel: 40 79 30 00).

The Mineralogical Gallery of the Université Pierre and Marie Curie is open every day, except Tuesdays, from 1pm to 6pm (34, rue Jussieu).

On the top floor there is a restaurant open to the public from which there is a panoramic view of the Seine, the bridges and the monuments of Paris. The Institut du Monde Arabe is open for guided visits led by specialised guides[1].

On the very wide banks of the Quai Saint-Bernard, in the seventies and the eighties, the Paris City Authorities had a promenade garden laid out, designed by the architect Daniel Badani: there is also an open air museum of sculptures with works from the great contemporary artists, such as Marino Di Téana, Marta Pan, Nicolas Schoeffer, Augustin Cardenas, Michel Guino, Michaël Noble, Maurice Lipsi, César, François Sthaly, Marta Calvin, Ipoustéguy, Brice, Reinhoud, Libéraki, Parvine Curie, Osip Zadkine, Bernard Pagès, Anthony Caro, Rougemont, Pierre Sabatier etc...

1. Entrance: 1, Rue des Fossés St-Bernard (Tel: 40 51 38 38). Visits every day, except Mondays from 1pm to 8pm. Exhibition room open every day, except Sundays and Mondays, from 10am to 6pm.

The Institut du Monde Arabe, at the corner of the Quai Saint-Bernard, facing the Pont Sully.

The Quai Saint-Bernard as seen from the dock of the Arsenal in 1864. A painting by Stanislas Lépine.

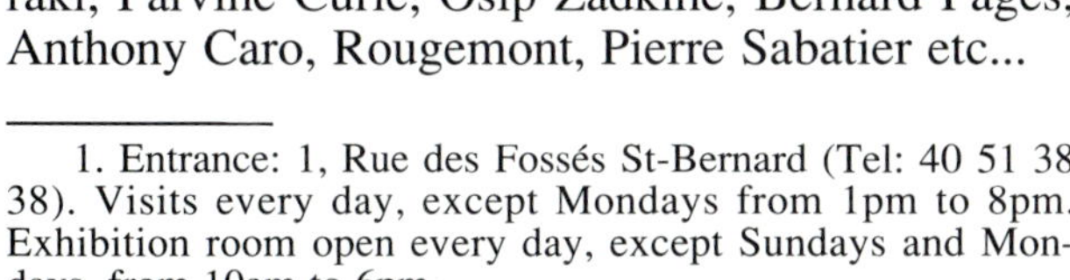

The quai Henri-IV

Begins at the Pont Morland
and ends at the Pont Sully.

The lock of the Arsenal.

This quay, which has been in existence since the Louis-Philippe period, was inaugurated on 7th July 1843.

The ground between the Quai Henri IV and the Boulevard Morland formed, before 1843, the island «île Louviers». This island - whose perimeter is today defined by the Boulevard Morland and the Quai de la Rapée, joined to the right bank in 1843 - was, for a long time, a pasture. Acquired by the City in 1671, it became a firewood «dépot» that Turgot transformed in 1737 into a lumberyard - planks, beams, thick boards, billets - for the craft industry and the building trade. The ancient prints which represent it (opposite) show a multitude of stacks of lumber, skilfully ordered to dry evenly: they were called «théatres».

The island owed its name to Nicholas Louviers, Seigneur de Cannes, Prévôt des Marchands (Provost of the Merchants), who was its owner in the middle of the XVth century.

On the Quai Henri IV are some buildings of neo-classical architecture. N°30 is an elegant building with two symmetrical wings flanking a courtyard, built in 1875-80 to house the Archives of the Seine department. N°s 32-34-36 are decorated with sculptures.

The high tower, intended to house the engineering departments of the City, is a particularly disastrous achievement of the sixties both for its aesthetic and visual impacts on the environment.

N° 38 is outstanding with its neo-Louis XIII architecture, its perfect symmetry and the six fine stone ram heads which support the balcony.

In 1992, the quay was completely re-arranged by the «Direction de la Voirie» of the City of Paris (the Transport Department); a promenade was created above the expressway, overlooking the Seine; benches and moulded iron lights were placed there. We are led to believe that this elegant solution will be applied to other parts of the quays of Paris.

Building n° 38, quai Henri IV, detail of balcony.

The Big Pier

As early as 1750, the upstream mouth of the right bank (or Notre Dame) arm of the Seine was blocked by a pier on piles between the tip of the Ile Saint-Louis and the Ile Louviers which served essentially as a timber yard.

This part of the river was then transformed into a marina and unloading port, well sheltered from high water and floating ice.

The pier had an opening to allow the passage of convoys, which was blocked, when the water began to freeze, by timber beams or by a sacrificed ship placed crosswise. This pier, which figures on a great number of views of Paris, drawings and paintings, was rebuilt several times, notably in 1818 and 1837. At that date, it was combined with a foot bridge which spared pedestrians the detour by the Pont Marie. It was called the Damiette foot-bridge.

The big pier was largely destroyed by the 1910 flood. It was rebuilt again from iron girders on which the foot-bridge was replaced. It was only in 1932 that the pier finally disappeared. At that time it served only to protect the last wash houses at the Pont Marie.

The big pier, circa 1900. A water-colour by René Leverd.

The pont Sully

The bridge or more precisely the bridges, because there are two separate bridges which link the Boulevard Henri IV and the Boulevard Saint-Germain, were built in 1874-1876. These iron bridges were achieved by the engineers Vaudrey and Brosselin.

They are 20 metres wide between their moulded iron parapets. The «left bank» - big arm - bridge has three elliptical iron arches; the central arch spans 49.5 metres and the side arches 15.3 metres.

The «right bank» - small arm - bridge has a central iron arch, with a span of 42.3 metres, the two semicircular side arches are in masonry and span 15.3 metres. Each arch has eleven iron trusses.

These bridges are not perpendicular to the banks but form an angle of about 45 degrees with them. The piers and abutments, placed in the axis of the current, are oblique, which gives an interesting geometry to these horizontally offset iron arches. The masonry piers, measured out with bosses, are protected by semicircular pier heads. The balustrades, with their very conservative outline, are made of moulded iron.

The square at the tip of the island and the central reservation between the two bridges, Ponts Sully, formed, in the XVIIth and XVIIIth centuries, a part of the garden «à la française» of the Hôtel de Bretonvilliers, destroyed in 1840.

The Seine, the Pont Sully and Notre-Dame.

The Pont Sully as seen from the Quai Henri IV.

The Pont Sully, part of the big arm on the right bank.

The Days of the Floating Wash Houses

Since the origins of the city, laundering had been done on the banks of the Seine. It was only at the end of the Middle Ages, when Paris became a large town, that the first floating wash houses appeared and that laundering was controlled by a real corporation.

The first wash houses were simple «punts» or flat-bottomed «barges» set up in places designated by the Provost of Paris, and more or less sheltered from bad weather by thatched roofs or weather boarding.

The places were granted to poor people by the town council authorities «so that they should earn their crust in this industry».

In 1789, there were about eighty wash houses, mostly placed along the best exposed parts of the right bank.

In 1796, the first industrial laundry was set up downstream of the former Ile des Cygnes in the plain of Grenelle.

In 1805, an Imperial order cancelled all further permits for the setting up of «laundry boats» which were considered as a hindrance to river traffic.

However, with the Restoration and the Monarchy of Louis-Philippe, the number of floating wash houses in Paris increased again. To fight against the competition from the laundries, the capacity and comfort of these establishments were improved. They were designed in imitation of floating public baths and swimming schools, very fashionable at the time, such as the Viguier and Deligny public baths. Alongside the wash houses were more archaic and rudimentary pontoons. These created a colourful scenery as can be seen in a splendid lithograph by Franck Boggs picturing the Quai de Montebello, where one can see a large number of poles on which the many washerwomen positioned just opposite, under the wall of the garden of the Archbishop's Palace, dried their laundry.

Floating wash houses near the Pont d'Arcole, circa 1885. In the foreground a steam passenger slow boat.

The large wash houses, true floating towns several dozen metres long, increased in number in the middle of the XIXth century, their favourite sites remaining the sunny banks along the right bank, from the Pont Marie up to Chaillot.

With the invention of steam boilers during the reign of Louis-Philippe, the wash houses were generally built on two floors. The laundresses worked at water-level, protected from the rain and sun by awnings: on the second floor were the large drying rooms.

At the end of the XIXth century the biggest and most prestigious of the wash houses was the «Arche Marion», a sort of two hundred metre-long liner between the Pont d'Arcole and the Pont Notre Dame. Richly decorated, it consisted of twelve barges. Two hundred and fifty washerwomen worked on this «flagship» of the Parisian wash houses. It disappeared at the end of the Belle Epoque.

The last floating wash houses of the Pont Marie stopped working around 1937.

A floating wash house on the Quai de la Mégisserie, near the Pont Neuf, at the beginning of the XIXth century.

The Raft Wood Ports

For many centuries, Paris was heated by wood. From the end of the XVIth century, in order to preserve the forests of the Ile-de-France, it was necessary to fetch firewood from great distances. The floating of timber rafts from Morvan down to Paris became quite usual from 1560 onwards.

Before this period, the principal users of logs and faggots were the bakers, the pastry cooks, the sellers of roast meats etc... The wood transported by boat from Brie, Champagne and near Burgundy, used to arrive at the «log port». The Rue de la Bûcherie, opened in the XIIIth century, serves as a souvenir of these activities.

After the beginning of «floating», the log port was moved towards the Quai Saint-Bernard and the Quai de la Tournelle, and later on, to the Ile Louviers (1), the Quai de la Rapée and the Quai de Bercy.

Downstream, the biggest wood port was that of La Grenouillère, between the Rue du Bac and the Pont de la Concorde.

Further down, the Ile Maquerelle was used to «rip up» the boats which were resold as firewood. In reality most boats which arrived at the Port of Paris only ever made one voyage.

The breaking up or «ripping up» was done by hand. About twenty men were needed to «rip up» a timber raft rapidly. In 1824, 1,137,000 steres of logs, 1,316,000 faggots and 1,690,000 bundles of kindling were unloaded.

These impressive quantities of firewood were supplied not only to the public, but also to bakers, dyers, brewers, the potteries and various other industries.

Floating by raft, an economical and convenient process, was invented in 1546 by a master smithy from Château-Chinon. The rafts measured from 72 to 75 metres long, and were 5 metres wide. Each could carry up to about two hundred steres.

Once loaded, several rafts were simultaneously launched on the river to take advantage of the flow caused by the opening of the floodgates of the Clamecy barrier, the starting point for the timber rafts.

The navigation of the timber rafts was dangerous and difficult, and demanded experienced «raftsmen». Only two men, the raftsman and his apprentice, had the formidable task of bringing the «raft» from Morvan right down to the Paris waiting ports. It took about ten days to come within view of Maisons-Alfort, Ivry and Charanton. There, the raftsman would take a «pilot» who would help him to reach the breaking-up ports. Sometimes it was necessary to wait for weeks, even months, to be able to approach the port or to break up. The raftsman and his apprentice would stay on their raft. Once the cargo was delivered to the consignee, the raftsmen would return to Morvan on foot. It was like this for about four hundred years between 1547, the date of the first convoy, and 1927, when the last rafts left the meadows of the Ile Margaut in Clamecy.

The breaking up.

The Ile Louviers and the «theatres» of drying timber. A print from the middle of the VXIIIth century.

The quai des Célestins

Begins at the Pont Sully
and ends at the Pont Marie.

This quay overlooks the ancient Port of the Célestins or Port Saint Paul. Right up to the 1840s it was the terminus for the barges on the Seine, which were replaced in about 1830 by steamboats. Its name comes from the presence of the ancient Convent of the Célestins, previously situated nearby, and destroyed during the French Revolution (Boulevard Henri IV).

Upstream, the first houses on this quay face the public garden Henri Galli where, in 1909, the foundations of one of the towers of the Bastille were rebuilt with blocks found during the construction of the Metro.

The mansion, the Hôtel de Fieubet, at n° 2, on the corner of the quay and the Rue du Petit Musc, is outstanding with its abundant sculpted decoration, high relief garlands of fruits and flowers, caryatids and telamones.

The Quai des Célestins and the Estacade in 1870. A painting by Stanislas Lépine.

The Hôtel Fieubert (Ecole Massillon) on the Quai des Célestins. A drawing by Fraipont, 1890.

Anne de Fieubet, Parliamentary advisor and Master of Petitions, had this mansion built by Jules-Hardouin Mansart. Mansart's design was altered in the XIXth century, in particular by the addition of a bell tower and belvedere. In 1877 the Oratorians installed the Massillon College which is still there.

The group of buildings which hides the Hôtel de Sens was rebuilt in the 1950s. This mansion is the most outstanding monument on the Quai des Célestins. It houses the Forney library (decorative arts, arts and crafts). It is one of the oldest constructions in Paris. Its Gothic portal, placed between two turrets in the form of watch towers with pepper pot like roofs, immediately draws up an image of the Middle Ages.

Up to 1622, the diocese of Paris, was a bishopric dependent on the arch bishopric of Sens; the archbishops of Sens, having more business in Paris, the seat of power, than in Sens, lived essentially in the capital. Their town house was built between 1475 and 1507 by the archbishop Tristan de Salazar.

The main façade, on the Seine side, on the corner of the Rue du Figuier is characterised by a Gothic portal with a diagonally-ribbed arcade flanked by two side doors. The building which

The Hôtel Royal of Saint-Pol

By locating the seat of the Royalty in the district of Saint-Paul and by making the mansion he had just had built there his main residence, Charles V broke with the ancient tradition that the kings and the court should reside in the Palace of the Cité or in the Louvres whenever they were in Paris, which was fairly infrequent.

The Hôtel Royal de Saint-Pol stretched from the Port Saint-Paul to the Rue Saint-Antoine. It was a rather ill-matched collection of buildings and gardens. The main entrance was towards the river. Strong walls surrounded the perimeter. This palace with its many rooms, its state apartments and outbuildings, formed a veritable district. Today only a few parts remain in the Rue Charles V and the Rue des Lions, a name which calls to mind the ancient royal menagerie.

The Hôtel de Sens circa 1920. A painting.

today houses the reading rooms and the stocks of the Forney library consists of a ground floor and a first floor crowned by a large gable. All the buildings have undergone major restoration, certain parts having been rebuilt.

The numbering of the buildings in the streets of Paris is further evidence of the close relationship between the city and its river.
The Préfet Frodiot established this numbering system. From 1805, the streets, avenues and boulevards were classified in roadways parallel or perpendicular to the Seine. The numbers increase from upstream to downstream for the roadways parallel to the river - the quays, the rue de Rivoli, the boulevard Saint-Germain for instance - and also in increasing numbers from the centre to the outskirts on the right bank, and on the left bank from the quays, the even numbers being on the right and the odd numbers on the left. The numbering of the quays of the Ile de la Cité and the Ile Saint-Louis also follows the direction of the current, from upstream to downstream.

The small arm of the Seine as seen from the quai des Célestins.

The quai de l'Hôtel-de-Ville

Begins at the Pont Marie
and ends at the Pont d'Arcole.

In the past, lower downstream, were the Port de la Grève and the Port des Ormes or Saint Paul; these were timber, hay and corn ports, and, from Autumn to the end of Winter, the floating apple port.

The modern quay encompasses the ancient Quai de la Grève which existed as early as the XIIIth century, as a well laid-out, slightly sloping bank .

The ancient houses have almost all disappeared except those on the south flank of the Saint-Gervais Church .

The «Cité Internationale des Arts» (International City of Arts) was built by the architect Olivier-Clément Cacoub in 1965.

The French garden of the Hôtel d'Aumont stretches out between this building and the Hôtel d'Aumont. Now the seat of the Tribunal Administratif of Paris, the Hôtel d'Aumont was built between 1644 and 1648 by the architects Louis Le Vau and F. Mansart. The main building has 2 rows of 17 bay-windows on the Seine side. A gable «à la française» crowns this very elegant façade.

The Hôtel d'Aumont.

A row of tall houses with unequal roofs, is all that is left of the former Quai des Ormes, today Quai de l'Hôtel-de-Ville. Beyond the houses, the roof and steeple of the Saint-Gervais Church can be seen. For a few dozen metres, the scene has not changed for two centuries, whereas in the east and the west, demolition has caused the picturesque rue de l'Hôtel-de-Ville to disappear.
The Rue de l'Hôtel-de-Ville, where most of the houses were pulled down during the 1920s, used to be called the Rue de la Mortelerie, because of the high population of plasterers who lived there.
The Saint-Gervais Saint-Protou Church was begun at the end of the XVth century and was only finished a century and a half later, which explains the contrast between the chevet, with its flamboyant Gothic style, and the façade achieved between 1616 and 1657 by the architects Salomon de Brosses and Clément Métézeau.
It houses a famous organ, erected between 1758 and 1768, by the organ builder Cliquot for the composer François Couperin.

The corner of the Quai de l'Hôtel-de-Ville and the Rue du Pont Louis-Philippe.

Up to the 1830s, the «Hôtel-de-Ville» (Town Hall) of Paris was still roughly as it had been built three centuries earlier, on the plans of the architect Dominique de Cortone, called the «Boccador». The Préfet Rambuteau had it enlarged in 1840 keeping the Renaissance style part. The inside was decorated by Henri Lehmann, Auguste Hesse, Bénouville, Cabanel, Desgoffes, Delacroix, who ended their works under the Second Empire. The entirety of the works of art and records were lost in the burning of the building by the Commune in May 1871; the fire lasted 8 days.

The rebuilding, entrusted to the architects Ballu and Deperthes, was achieved between 1874 and 1882. Over 300 sculptors, painters and decorators contributed to the decorating of the grand staircases, the many sumptuous state rooms and the reception room, situated on the Rue Lobau side, above the Saint-Jean room.

The statues of 196 famous characters were placed in the niches of the façades. The painted and sculpted ornaments of the Hôtel-de-Ville, a true

The Quai de l'Hôtel-de-Ville, the Port Saint-Paul, in the background the Pont Notre-Dame and its houses. Print by Lespinasse, middle of the XVIIIth century, Musée Carnavalet.

The Hôtel-de-Ville and the quays on the right bank at the beginning of the Second Empire.

urban castle where the neo-Renaissance style is associated to the neo-classical style, represented an important part of the total cost: 8.4 million out of a budget of 30.5 million francs.

On the Square of the Hôtel-de-Ville, the two large bronze statues represent «Science» by Jules Blanchard (on the left) and «Art» by Laurent Marqueste (on the right). On either side of the clock, the stone groups represent «Work» by Ernest Hiolle, «the Seine and the Marne» by Aimé Millet, «the City of Paris» (above the dial) by Jean Gautherin, and on the pediment «Prudence» and «Vigilance» by Charles Gauthier. The large gables «à la française» dominated by the high bell tower, are crowned with six chased copper statues, overlaid with gold, representing horsemen of the Middle Ages leaning on their banners.

On the Seine side, the bronze equestrian statue representing Etienne Marcel, Provost of the Merchants of Paris (the Dean of the Guild) in the XIVth century is the work of the sculptors Jean Idrac and Laurent Marqueste . It was inaugurated in 1888.

Among the most outstanding sculpted elements on each side of the portals which open onto the Rue Lobau, are four big bronze lions by Alfred Jacquemard and Auguste Cain. On the upper part of this façade, fourteen allegorical figures represent the major cities of France.

The statue of Etienne Marcel, the Garden of the Hôtel-de-Ville. A water-colour by René Leverd, 1900.

The port and the place de Grève

Before the XIth century, the Place de Grève was a wide, low bank sloping slightly towards the Seine. In 1141, Louis VII the Young handed this space over to the middle-classes of the Hanse, the merchants on the water, who had the monopoly over the river traffic on the Seine and its tributaries. It became a sort of annexe for the Port Saint-Landry which, narrow and congested, was opposite, on the North flank of the Ile de la Cité.

The Port de la Grève developed rapidly and was actually made up of several more or less specialised ports along the right bank between the present Pont Louis-Philippe and Pont au Change: on that site there were hay, wine, timber, salt and cereal ports (Rue du Grenier sur l'Eau), and many mills.

The activity of the Port de Grève only ceased in the XIXth century with the construction of quays. The port activities continued on the low banks, with in particular the apple port: every Winter barges fully laden with apples would moor and the fruiterers of Paris would come and buy there (see painting).

The first suspended foot-bridge (the Pont d'Arcole) was built, in 1828, facing the Place de Grève.

Created in 1246 by Saint-Louis, the town administration, a group of aldermen presided over by the Dean of the Guild, used the «parloir aux Bourgeois» near the Grand Châtelet as a

The Place and the Quai de Grève and the Hôtel-de-Ville at the beginning of the XVIIIth century. A print by Rigaud.

The Place de Grève and the Pillared House at the end of the Middle Ages. A reconstitution by Hauffbauer.

meeting place and town hall for a long time. In 1357, Etienne Marcel, the Dean of the Guild, a turbulent but active man, transferred the town administration to a house on the Place de Grève called «Maison aux Piliers» (the Pillared House).

Under the reigns of François I and Henri II, during the Renaissance, then at the beginning of the XVIIth century, in the days of Henri IV and during the youth of Louis XIII, a large town hall was built on the plans of the master mason and architect Dominique de Cortone, known as Le Boccador. Pierre Chambiges, a master mason and builder took part in this construction which existed, among the buildings erected under Louis-Philippe, right up to the great fire of the Commune in May 1871.

The Place de Grève was much smaller than the Place de l'Hôtel-de-Ville which took its definitive layout under the Second Empire with the opening of the Rue de Rivoli, of the Avenue Victoria, and the construction of two buildings on either side of this avenue.

A beheading Place de Grève, circa 1740. A reconstitution by Hauffbauer.

The apple market on the bank of the quai de l'Hôtel-de-Ville, circa 1890. A lithograph by Fraipont.

The Place de Grève was a place of festivities, receptions, fireworks and countless events. It was officially called Place de l'Hôtel-de-Ville in 1802, nevertheless the former name prevailed throughout the XIXth century.

After the enlargements by Haussmann, and during the first reconstruction of the Hôtel-de-Ville under Louis-Philippe, the square was used each morning as a gathering place for the masons of the Creuse department and other building workers, numerous in this district, who used to come there to be hired by architects, foremen and master builders: it was called «faire la grève» (cf. Martin Nadeau, Les Mémoires de Léonard - ancien garçon maçon, a former mason's boy). The expression «faire la grève» has been retained but it now has a totally opposite meaning...

The Hôtel-de-Ville and the old Pont d'Arcole, circa 1846.

The Hôtel-de-Ville and the Pont d'Arcole in 1900.

The Hôtel-de-Ville today.

The quai de Béthune

Begins at the Pont Sully
and ends at the Pont de la Tournelle.

This quay was for a long time called the «Quai des Balcons». In 1806, it received the name of Maximilien de Béthune, Duc de Sully (1560-1641), principal minister of Henri IV.

The name «Quai des Balcons» originated from a suggestion by the architect Louis Le Vau that all the mansions of the quays of the Ile Saint-Louis should be decorated with balconies.

The buildings which occupy the space opposite the Boulevard Henri IV, between the two parts of the Pont Sully, are on the site of the magnificent Hôtel de Bretonvilliers. Built between 1637 and 1640, following the plans of the architect Jean Andronet du Cerceau, for Claude Le Rageois de Bretonvilliers, Secretary to the Council. This mansion disappeared in 1840.

The chronicler, Tallemant des Réaux, wrote that this mansion and its garden formed the best situated combination in the world after the Seraglio of Byzantium...

The garden, which is on the upstream tip of the island, occupies the site of the garden of the Bretonvilliers. The pedestal of the statue of the sculptor Louis Barye, the famous animal sculptor of the end of the XIXth century, can be seen there. This effigy, by Laurent Marqueste, melted down during the Occupation has not been replaced. The two stone groups of the pedestal are decorated with beautiful lions, also the work of Laurent Marqueste.

The quai de Béthune

Buildings on the Quai de Béthune.

Famous people on the Quai de Béthune.

The Rue de Bretonvilliers, between the Quai and the Rue Saint-Louis-en-l'Ile, passes under a three-storied building, the last of some other buildings the Marquis de Bretonvilliers had built.

At n° 18, Le Vau erected for Thomas de Coomans d'Astry, a mansion which was the property, in the XVIIIth century, of the Duc Louis-François Armand de Richelieu, Marshal of France. It has a superb entrance, with columns and arcades leading to a monumental staircase. The great courtyard is imposing with its double tier of pilasters.

N° 20 and n° 22 are the town houses «Le Febvre de la Barre» and «Le Febvre de Malmaison». N° 20 has a beautiful staircase and a drawing-room which still has its original ceiling painted by Mignard. N° 22 is similar to the preceding one, also with a beautiful staircase. The poet Francis Carco, who wrote about thirty realistic

Pavilion of the Hôtel Bretonvilliers. Portico of the rue Saint-Louis-en-l'Ile.

narratives and novels about Paris, lived in this house for a long time, and before him, Charles Baudelaire.

One of the mansions which figured among the masterpieces of the isle was replaced in the 1930s by the present n°24. This mansion had been built for Louis Hasselin, Master of the Chamber of Deniers. The portal, whose panels were sculpted by Etienne Le Hongreis, is the sole remaining part. President Georges Pompidou lived in this building.

René Cassini (1887-1976), winner of the Nobel Peace prize, lived at n° 36, as did Marie Sklodowska (1867-1934), born in Warsaw, better known under the name of Marie Curie.

The Hôtel de Bretonvilliers upstream of the Ile Saint-Louis, in the middle of the XVIIIth century. In the background the Hôtel Lambert. Painting by Raguenet, Musée Carnavalet.

The quai d'Orléans

Begins at the Pont de la Tournelle
and ends at the Saint-Louis foot-bridge.

Facing south-west, towards the chevet of Notre Dame and the West bank, quite far from the heavy traffic, this quay has an exceptional location. It bears the name of Gaston d'Orléans, brother of Louis XIII.

At n° 6, in a town house built in 1655 there is the library[1] and the «Adam Mickiewicz» Polish museum, founded in 1903 by Ladislas Mickiewicz, son of the Polish patriot poet exiled to Paris. There are here, many documents about the Romantic period, souvenirs of Frédéric Chopin, etc.

The poet and writer Jean de la Ville de Mirmont (1886-1914) lived at n°8.

N°12 catches the eye with its splendid undulating wrought-iron balcony. It was built by Michel Guillaume who, in the 1640s, had most of the houses on this quay built.

At the corner of the quay and the Rue Guillaume Budé a high bronze relief represents the profile of the poet Félix Arvers.

The Rolland town house occupies n°18 and n°20: an inscription serves as a reminder that this house was the property, in 1775, of Etienne François de Turgot, Marquis de Saumont, governor of French Guyana.

At the corner of the Rue Boutarel an inscription testifies to the memory of the painter and engraver André Dignimont who illustrated, in particular, the works of poets and writers such as Francis Carco, Pierre Mac-Orlan, André Salmon, Roland Dorgelès, Alexandre Arnoux, Paul Fort, etc ...

Town houses on the Quai d'Orléans.

The Polish Library.

1. The library is open from Tuesday to Friday from 2pm to 6pm. The museum on Thursdays only from 3pm to 6pm.

The quai d'Anjou

Begins at the Pont Sully
and ends at the Pont Marie.

This quay was built, by the master builders Christophe Marie, Le Regrattier and Poulletier, between 1615 and 1650 during the parcelling out of the land.

On the corner of the quay and Rue Saint-Louis-en-l'Ile, there is the Hôtel Lambert, built by the architect Louis Le Vau between 1642 and 1644. It is a magnificent residence, built around a main courtyard and a terraced garden. The façade overlooking the garden has two wings; it ends with a rotunda. The main wing contains the famous «Galerie d'Hercule» whose great vaulted ceiling was painted by Charles Le Brun.

This town house was built for Jean-Baptiste Lambert de Thorigny, a counsellor and secretary to the King. Apart from Le Brun for the Hercules gallery, Eustache Le Sueur, François Perrier, Pier-

The Hôtel Lambert and the Quai d'Anjou.

re Patel and other artists took part in the decoratien. In the XVIIIth century the town house belonged to Claude Dupin. His wife, the daughter of the financier Samuel Bernard, a friend and custodian of Jean-Jacques Rousseau, ran a very well esteemed salon there.

This town house went through diverse fortunes before being bought by Prince Adam Czartorisky who undertook its complete restoration. The Czartorisky princes kept and maintained this masterpiece of the French decorative and architectural heritage. Baron Guy de Rotschild, who has since acquired, has continued this tradition.

The semi-detached building at n°3, which shares its party wall with the Hôtel Lambert, was also built by Louis Le Vau, who himself lived there.

N° 5 was the property of the Marquis de Marigny, «Directeur Général des Bâtiments du Roy». Rennequin Sualem, designer-builder of the Machines of Marly which raised the Seine water to the gardens of Versailles, lived there.

N° 7, built in 1642 for Jacques Brébert, an iron merchant, has housed the Syndicat Patronal des Boulangers et Pâtissiers (the Masters' Federation of Bakers and Pastry-cooks) since 1843.

The buildings, numbers 9, 11, 13, 15, of great similarity, were built in 1641-42 for Jean-Baptiste Lambert de Thorigny by Louis Le Vau. The artist Honoré Daumier lived at n°9 from 1846 to 1863. Charles Daubigny, one of the great landscape-painters of the XIXth century had his studio at n° 13, as did the sculptor Geoffroy-Dechaume.

N° 17 is the Hôtel Lauzun, today property of the Ville de Paris. It was built by Louis Le Vau in the 1650-60s, for Charles Gruyn des Bordes. In 1682, it became the property of Antoine Nompar de Caumont, Duc de Lauzun, from whom it takes its name. It then passed successively to the families of the Duc de Richelieu, the Marquis de Pimodan in the XIXth century, and eventually to the Baron Jérôme Pichon, who had it restored. Before, this town house had been rented to a certain number of men of letters and artists, such as

The Hôtel of Lauzun

The interior decoration is of exceptional quality. The ceiling of the grand staircase is painted. It is an allegory attributed to Le Brun. Two beautiful statues, Minerva and Apollo, are placed in niches. Above the doors, painted friezes symbolise Poetry, Music, the Arts and the Sciences.
The galleries are decorated in an incredibly luxurious style, typical of the Louis XIV period. The over-mantles, overdoors, splays, ceilings, cornices and panelling are the work of the best woodcarvers, staff makers, decorators, painters and gilders.
The grand bedroom has a painted ceiling along the theme of the «Triumph of Venus».
The sumptuous decoration is epitomised in the room called «little boudoir». The mirror effects give infinite reflections of the multicoloured richness of the walls and ceiling, the central theme of which - an allegory of Flora and Zephyr - is treated with a rare exuberance.

The quai d'Anjou by Leverd.

Roger de Beauvoir and Charles Baudelaire who wrote the first pages of the «Fleurs du Mal» there. There, Baudelaire and Roger de Beauvoir would receive and entertain all their rowdy friends; writers, artists, actresses, musicians. To a complaining neighbour, Baudelaire once answered: *«Sir, I chop wood in my drawing-room, I drag my mistress by her hair, it is done everywhere, and you have no right to interfere».*

The building situated at the corner of the quay and of the Rue Poulletier (entrance at n°20 of this street) is also an ancient XVIIth century town house. Recently restored, it is occupied by a primary school, whose classrooms partly overlook the Seine.

The Hôtel of Lauzun.

N° 23 was the home, in 1624, of Gabriel Sionite, an Arabic teacher and chief of the Maronites of Lebanon.

N° 27 was the home of Salomon de Croix, an architect of the Louis XIV period, and Nicolas Sainctat, the man who introduced Ambassadors.

In 1644, n°31 was the town house of the «Maître des Requêtes» Boucher d'Orsay - who was at the origin of the creation of the Quai d'Orsay.

N° 35, Quai d'Anjou, was, in the middle of the XVIIIth century, the property of César François Cassini de Thury (1714-1784) who began drawing a great map of France, the «Cassini map», which was finished by his son Dominique. This immense work was the origin of the division of the country into departments, under the Convention. These two geographers were descendants of the astronomer Jean-Dominique Cassini (1625-1712) called to France by Colbert to organise the Observatory of Paris.

Most of the plaques commemorating these different personalities were mounted, in the 1930s, by the architect Joseph Marrast, whose family has lived on the Quai d'Orléans for over a century and a half (Armand Marrast, writer and journalist, the first elected mayor of Paris already lived on the Ile Saint-Louis). All these plaques bear, in a medallion, the effigy of Saint-Louis.

The balconies, quai d'Anjou.

The Hôtel du Syndicat de la Boulangerie et de la Pâtisserie, quai d'Anjou.

The quai de Bourbon

Begins at the Pont Marie
and forms the downstream tip of the isle,
down to the Pont Saint-Louis.

The sign of the Cabaret du Franc Pinot.

On the ground floor of the building, at n°1, there is the «Cabaret du Franc Pinot» (the pinot is a variety of black grape which gives the best Burgundy wines). Established in the XVIIth century, its appearance has remained unchanged. It was one of the starting points of the barges which navigated the Seine, and a meeting place for the bargees and true drinkers ...

The house at n°11 was built in the 1640s for the painter Philippe de Champaigne, court painter of Queen Marie de Médicis and portrait painter of Louis XIII, Richelieu and Anne d'Autriche.

At the bottom of the second courtyard there is the last «jeu de paume» (real tennis court) in Paris, built under the reign of Louis XIII; its entrance is at n°54 Rue Saint Louis en l'Ile.

The Hôtel Charron, at n°15, is a fine building with a pyramidal roof. The high windows on the ground floor, either side of the portal, still have their wrought-iron gratings. This town house was built for Jean Charron, a controller of the «Extraordinaries des Guerres en Picardie». The painter Emile Bernard, a great friend of Van Gogh, lived there in the XIXth century.

... In memory of a sculptress ...

The large building at n°19 is the Hôtel de Jassaud. It is the most beautiful residence on the Quai Bourbon, its façade, of over 30 metres, is embellished with floral pediments. The main balcony, with its elegant wrought-iron railing is supported by splendid consoles. This town house was built in 1666 for Nicolas de Jassaud, Maître des Requêtes under Louis XIV, and later Secretary of State. A recent plaque informs us that Camille Claudel, a sculptress (1864-1943), had a studio there between 1899 and 1913.

N°21, at the corner of the Rue Le Regrattier, was built, as were the next two houses, for Nicolas Gaillard de Pommeray, Auditeur des Comptes

(Commissioner of Audit). It is a fine house with no exceptional exterior features, but some of its rooms were painted by the pupils of J.F. Boucher. In a niche, a mutilated statue explains the former name of the Rue Le Regrattier - formerly street of the Headless Woman.

The town house at n°29 is outstanding out with its fine carriage entrance and fanlight. In the XVIIIth century, the Counsellor Louis Roualle de Boisgelin lived in the town house to which he left his name. At the end of the XIXth century, Emmanuel Lansyer, a pupil of Courbet, had a studio there. Emile Zola found his inspiration from the artists' studios on the Quai Bourbon, which he described magnificently in his novel «The Masterpiece».

On the façade of n° 31 two plaques honour the memory of Théophraste Renaudot (1586-1653), the founder, in 1631, of the newspaper «La Gazette» and the writer Charles-Louis-Philippe (1874-1909).

At n° 45 a plaque tells that Princess Bibesco held her literary salon there. N°49 is called the «Maison du Centaure» (Centaur's House) because of the presence of two high relieves representing Hercules armed with a club to kill Nessus. Guillaume Apollinaire, in particular, lived in this building.

The painter Jean-Baptiste de Champaigne, a nephew of Philippe de Champaigne and decorator of the Saint-Louis-en-l'Ile Church, lived at n°51.

The Rue Jean Du Bellay.

A high relief of the Maison du Centaure.

Floating wash-houses Quai de Bourbon, circa 1920.

The pont Marie

Built between 1614 and 1635 by Christophe Marie and Jean Delagrange, this bridge, whose superstructure has been altered, consists of five semicircular arches of unequal span, resting on 4-metre thick piers.

Its width of 22 metres, exceptional for the time, can be explained by the fact that the Pont Marie was designed to bear two rows of houses. These were destroyed in 1786. The work is 92 metres long. Above the triangular pier heads, 8 fine cul-de-four niches, decorated with pediments and pilasters have never received the statues for which they were designed.

The Pont Marie, 1880. A drawing by Fraipont.

A floating wash-house near the Pont Marie circa 1886. A painting by Gustave Maincent.

The pont de la Tournelle

Initially in wood, the Pont de la Tournelle, was built in the years 1621-1623, during the parcelling out of the land on the Ile Saint-Louis. Swept away by the waters in 1637, it was rebuilt in wood and destroyed once more in 1651. Reconstruction was started again in 1654. The bridge existed until the 1930s, having been altered in 1845-1846. The bridge had six semicircular arches with spans of 17.5 and 15.5 metres. The piers were protected by triangular pierheads.

During the reign of Louis Philippe, in 1845-46, the roadway was lowered and the superstructure widened with iron arches which bore the footways.

The pont de la Tournelle

This bridge, whose narrow arches hindered navigation, was replaced in 1928 by the present work, made of stone-clad reinforced concrete. It has a big depressed central arch and two semicircular side arches. It is 120 metres long and 23 metres wide.

A statue of Sainte Geneviève, by Paul Landowski, was placed upstream, on a white stone pylon.

The former pont de la Tournelle, circa 1900. Water-colour by René Leverd.

The pont Louis-Philippe

An iron suspension bridge was built in 1833 by the Seguin brothers to link the Quai de l'Hôtel-de-Ville to the Ile Saint-Louis and the Cité. It was called «Pont Louis-Philippe». The chains bearing the decks were suspended from the top of a high archway built on the west tip of the Ile Saint-Louis. It was reserved for pedestrians and light loads.

A new work, designed by the engineers Romany and Savarin, was undertaken in 1860-62. On this occasion, the Rue des Deux-Ponts was opened, thus allowing access to the Pont Saint-Louis.

The Pont Louis-Philippe, the Quai de l'Hôtel-de-Ville and the Saint Gervais Church as seen from the Quai de Bourbon. Painting by Serge Belloni, 1972.

The Pont Louis-Philippe and the Panthéon in perspective from the Rue Jean Du Bellay.

The Pont Louis-Philippe has three elliptical arches which are supported by 4-metre thick piers. In the middle, the «navigation» arch spans 32 metres; the side arches span 30 metres. The work is 15 metres wide and 100 metres long. At the keystone, the coarse-grained limestone vaults are one metre thick. The cornices, parapets, balustradcs and decorative occulis are made of stone. The piers, protected by semicircular pierheads are in stonework with bosses.

The pont Saint-Louis

The Quai, Pont de la Tournelle and the Pont Saint-Louis at the end of the XVIIth century. Painting by Théodore Matham, Musée Carnavalet.

Linking the Ile de la Cité to the Ile Saint-Louis, the Pont Saint-Louis was rebuilt in 1970, to replace a temporary passenger bridge set up in 1941. It consists of a single iron arch, with no aesthetic qualities.

The first wooden bridge linking the Cité to the Ile Saint-Louis was built in 1627, during the parcelling out of the land on the Ile Saint-Louis.

Rebuilt in 1710, it was called the «Pont Rouge» because of the red lead paint with which it was painted. A new work was built in 1803, at the beginning of the Empire. It was washed away in 1811 and was replaced by a passenger bridge which in turn was replaced by a suspension bridge in 1848.

In 1861, a new 16-metre wide iron work was built with a single arch spanning 64 metres, but this had to be pulled down in 1940 after being hit by a barge.

The Pont Saint-Louis, circa 1890. A drawing by Fraipont.

The quai de la Tournelle

Begins at the Pont Sully
and ends at the Pont de l'Archevêché.

The Musée de la Table,
Tour d'Argent restaurant.

This quay, built in 1340, used to overhang the Port de la Tournelle, the log and tile port. The first quay was built in 1554. It was given its present name in 1750. In the Middle Ages, «la Tournelle», the first tower of the Philippe Auguste wall, was built on this quay, on the left bank. Access to the Quai de la Tournelle was controlled by a gate which was rebuilt in 1606 and later replaced in 1670 by a veritable triumphal arch; built by the architect Jean-François Blondel. It was destroyed in 1787. There was also a château de la Tournelle where Monsieur Vincent, otherwise known as Saint Vincent de Paul (1581-1660) settled the convicts sentenced to row on the state galleys which he looked after. The Château de la Tournelle disappeared in 1790.

Opposite the Pont de la Tournelle at n°15, there is one of the most famous restaurants in the world, established way back in 1582, under the reign of Henri III. Even in those early days it was a very fashionable place, and legend has it that it was here, at the «Tour d'Argent», that the fork was first used. The founder of the dynasty of the «des Terrails» modernized it in 1914. It has remained in this family ever since. The top-floor panoramic restaurant has an exceptional view over Paris[1].

The Quai de la Tournelle as seen
from the Pont Saint-Louis in 1980.

N° 27 has a plain high quality façade. It was designed by the architect Leduc in the XVIIth century. It was at that time one of the town houses of the Clermont-Tonnerre family.

N° 37 is also a fine XVIIth century town house, transformed in its upper part; the inscription «Ci devant hôtel du Président Rolland» (former town house of President Rolland) has remained.

N° 47 was called «Hôtel de Miramion» because of the presence in the XVIIth century of the Filles de Sainte-Geneviève or Miramiones, an order founded by Madame de Miramion and which Madame de Maintenon and Louis XIV supported. In 1792, this town house was assigned to the central Pharmacies of the hospitals of Paris. The «Musée de l'Assistance Publique» was housed there in 1985. The most beautiful façade, flanked by two projecting parts, is on the garden side [2].

The adjoining town house, situated at the corner of the Quai and the Rue des Bernardins, also has an inscription dating from the Revolution «Hôtel ci devant de Nesmond» (former town house of Nesmond). It was built in 1636 by François Théodore de Nesmond, superintendant to the Prince de Condé. Today it contains the offices of the association «La Demeure Historique» which actively participates in promoting private monumental heritage.

The Rue de Bièvre which marks the end of the Quai de la Tournelle existed under that name in 1224. It was named after a channel dug in 1150 to bring the waters of this river to the gardens of Saint Victor Abbey. This channel was then used as a moat for the Philippe Auguste wall and the château de la Tournelle.

The private residence of the President of the French Republic, Mr François Mitterand can be found in this streat, which was transformed into a pedestrian precinct in 1981.

The Tour d'Argent and the Pont de la Tournelle.

1. A small «Musée de la Table» set up on the ground floor of the building is open to the establishment's clients. The Restaurant and Museum are open everyday except Mondays. Tel: 43 26 40 28 - 43 54 23 31.
2. Museum open everyday, except Sundays, Mondays and public holidays, from 10am to 5pm. Tel: 46 33 01 43.

The pont de l'Archevêché

This 68-metre long, 17-metre wide bridge links the left bank to the Quai de l'Archevêché (Square Jean XXIII), opposite the Rue de Bièvre and the Rue des Bernardins. It was built in 1828 by the engineer Plouard. In those days it ended near the Archbishop's palace in Paris. It is a three arch work in masonry, with 17 metre and 15 metre spans. The pierheads are semicircular. The balustrade is a simple lattice railing characteristic of the «Restoration» period. The Pont de l'Archevêché was a toll bridge, up to the Second Empire.

The Pont de l'Archevêché and Notre Dame in 1884. A painting by F. de Benoist.

The Pont de l'Archevêché, circa 1900. A painting by Franck Will.

The pont au Double

This bridge links the Quai de Montebello to the Parvis of Notre Dame. It takes its name from the toll of six deniers charged to pedestrians and a «double tournois» to horsemen. The toll was abolished in 1789.

Built for the first time in 1625 by the engineer Camard, it served mainly to link the two parts of the Hôtel-Dieu together, on either side of the small arm of the Seine where today there are the square and gardens of the Parvis de Notre-Dame, the Quai de Montebello and the Square Viviani. The Saint-Cosme wing of the Hôtel-Dieu was built on the bridge.

In 1634, the inhabitants of that part of the town obtained a right of passage on the bridge.

The present Pont au Double was built in 1847, by the engineers Bernard and Lax: it has a single iron arch with a 35 metre clear span. The superstructure is 45 metres long and 20 metres wide. The decoration of the work, on the bank arches as well as on the copper-covered moulded iron balustrades, has been well looked after.

The Statue of Charlemagne on the Place du Parvis Notre-Dame.

The quai de Montébello

Begins at the Pont de l'Archevêché
and ends at the Petit Pont.

The bank of the quay of Montebello, circa 1890. Painting by Firmin Girard.

This quay, which honours the memory of Maréchal Lannes, Duc de Montebello killed at Essling, is situated partly on the site of the ancient Pont aux Bûches and partly on that of the Hôtel-Dieu and of the Petit Châtelet.

The buildings of the Quai de Montebello were built at various periods between the XVIth and the end of the XIXth century. At the beginning of the XIXth century the houses were still often occupied by washerwomen who worked on the numerous floating wash houses and pontoons below the Quai de l'Archevêché.

The façades curve twice creating space for two charming little squares where, as soon as the weather is fine, the restaurants and cafés spread out their tables. It is the same at the corner of the

The Quai de Montebello and the floating wash houses, circa 1840. A print by Franck Boggs.

quay and the Rue Maître-Albert. At the corner of the Rue de Grands Degrés, named after a flight of steps which used to go down to the Seine, there are ancient advertisements painted on the wall. In the background of the small square formed by the widening of the Rue du Haut Pavé, there is an unusual view of the dome of the Panthéon.

The Rue de la Bûcherie serves as a backdrop for the Quai de Montebello between the Square Viviani and the Rue du Petit Pont. It is one of the oldest streets in Paris; it existed as early as the XIIIth century. It takes its name from the log port which was found there.

Maître-Albert, known as Albert Le Grand (1193-1280), a Dominican philosopher, one of the most famous scholars of the Middle Ages, taught at the University of Paris.

The beautiful building, topped with a high dome, located at the corner of the Rue de l'Hôtel Colbert and the Rue de la Bûcherie, is the surgery amphitheatre of the former Faculté de Médecine of Paris which was established on this site between 1472 and 1775.

The Hôtel-Dieu de Paris used to stand on the site of the present square, between the Pont aux Doubles and the Petit-Pont. Its disappearance, in 1878, opened up the panorama of the quays right back to the Saint-Julien le Pauvre Church and the Rue de la Bûcherie.

The origin of Saint-Julien le Pauvre - a plain church with neither steeple nor spire - goes back to the VIth century. The present building was constructed between 1170 and 1240. It is a building of the transitory period between the Romanesque and Gothic styles. It is also one of the three oldest churches in Paris, with Saint-Germain des Prés and Saint-Pierre de Montmartre.

N° 39 Rue de la Bûcherie, opposite the square, is, with its big single sloping roof and its single floor, one of the quaintest and smallest houses in Paris.

On the left there is the famous English library of Paris: Shakespeare and Company. On the right, there is the restaurant «La Bûcherie».

Houses on the Quai de Montebello.

The Petit-Pont

Between the Quai Saint-Michel and the Quai du Marché-Neuf, in line with the Rue Saint-Jacques, an ancient Roman road, the Petit-Pont, as its name suggests, is the shortest bridge in Paris (40 metres long). It has been rebuilt many times.

The Bishop of Paris, Maurice de Sully, responsable for the construction of Notre-Dame, had a masonry bridge built there in 1186; a bridge rebuilt in 1395, 1409, 1719. Permanently overcrowded, it was one of the rare passages to which all the highways converged.

The small arm of the Seine near Notre-Dame.

A toll was payable there ; the tumblers, bear-leaders and performing monkeys exhibitors heading for the royal palace of the Cité were exempted from paying, on condition that they made their animals perform a few tricks to prove that they were actually performing animals. This was known as «payer en monnaie de singe» (paying with monkey money).

The entrance of the Petit-Pont was on the left bank through the fortress of the Petit-Châtelet, which was destroyed after the Révolution.

For centuries, this bridge bore two rows of houses, but these were devastated by a fire in 1718.

In 1719 a three span stone bridge was built; it also bore two rows of houses, which disappeared in 1787-1788. Being an obstacle to navigation, this work was pulled down in 1852 and replaced by the existing single depressed arch bridge with a 32 metre span.

The arch and tympanum are made of burrstone, the end arches and the solid parapet are in ashlar. The work was achieved by the engineers de Lagalisserie and Darcel.

The Petit-Pont and the Petit-Châtelet at the beginning of the XVIIIth century.

The Petit-Pont, the Petit-Châtelet and the Hôtel-Dieu.

The quai Saint-Michel

Begins at the Petit-Pont
and ends at the Pont Saint-Michel

At the beginning of the Rue Saint-Jacques, the space between the quay and the Rue de la Bûcherie was occupied by the Petit-Châtelet. This fortress controlled the access to the Ile de la Cité, on the left bank, as early as the IXth century. It disappeared in 1782.

The construction of the Quai Saint-Michel, planned in 1561, then again in 1767, was in fact only started in 1812. Prior to that, the space was occupied by houses overhanging the river. Narrow streets used to wind down between these houses towards the Seine. Two have partly survived: the Rue du Chat-qui-Pêche (the Fishing Cat Street), the narrowest in Paris with its width of 2.50 metres, and the Rue Xavier Privas.

N° 19 is a fine building with semicircular arches from the Louis-Philippe period.

*The Quai Saint-Michel, circa 1890.
Painting by Galien-Laloue.*

The Quai and the Square Saint-Michel, circa 1890.
Painting by Galien Laloue.

N° 25 and n° 27 date back to the Second Empire; womens' heads can be seen in medallions in profile or on the keystones.

The place Saint-Michel, built in 1860 amidst a confusion of shacks and small streets which led doxn to the horses' drinking troughs on the banks of the Seine, was carefully designed by Davioud. The large, identical façades with their semicircular, ground level arches and their high windows with mens' and womens' heads carved into their keystones, give the square a monumental appearance. The tall Corinthian pilasters, present in all the squares designed by Hoffman, endow this square with a certain majesty.

The Saint-Michel fountain, also designed by Gabriel Davioud, was built in 1860 to provide a monument in the perspective from the boulevard du Palais and from the bridge.

The Saint-Michel square is the crossroads where generations of students and tourists have crossed.

One of the biggest and oldest bookshops in Paris, the Librairie Gilbert Jeune has been on the

The quai Saint-Michel.

square since 1971, having been located on the quay at n° 23 in 1890, and at n° 27 in 1909.

The founder of this dynasty of great bookshops, Joseph Gilbert, himself began as a bookseller on the quay in 1886.

The Saint-Michel Fountain : a detail.

This monument is typical of the utmost attention to detail that the Préfet Georges Eugène Haussmann and his assistants had for the quality of the streets and the urban aestheticism under the Second Empire. The Saint-Michel Fountain opens out in a space 26 metres high and 15 metres wide.

In the centre of the monument, Saint Michel the Archangel slaying the Dragon is a fine bronze work by the sculptor Francisque Duret. Four red marble columns with Corinthian capitals surround this beautiful central effigy. They bear the statues of the four cardinal virtues: Prudence, Fortitude, Justice and Temperance, by Jean Barre, Eugène Guillaume, Louis Robert and Charles Guméry. In the upper pediment two feminine allegories, by the sculptor Auguste le Bay, seated on either side of the coat of arms of Paris, represent Power and Moderation. The two big bronze dragons placed on the lip of the basin are by Henri Jacqueminot.

The Saint-Michel Fountain.

The Saint-Michel Fountain built by the architect Gabriel Davioud under the Second Empire.

The quai de l'Archevêché

Begins at the Pont de l'Archevêché
and ends at the Pont au Double.

The Deportation Memorial, the Square de l'Ile de France.

The Quai de l'Archevêché.

This quay merges with the public garden which can be found at the chevet, south of Notre-Dame.

At the upstream tip of the island there is another garden; the Square de l'Ile-de-France. In 1960-1961 the architect Georges-Henri Pingusson built the crypt of the Deportation Memorial there. In front is a space for meditation, surrounded by high walls and open at its lowest part to the waters of the Seine so that visitors, when there, can see only the water and the sky. A narrow passage leads down into this crypt and renders monumental the presence of two rough concrete sallies. This Memorial of Cistercian sobriety, by its strength and discretion, pays the most stirring homage to the martyrs of deportation. The tall, pointed steel sculpture symbolising the gates of the camps, was achieved by Roger Desesprit.

This space, which was in the past outside the wall of the Archevêché, is one of the only places on the Ile de la Cité on which there has never been any building. It was originally a separate islet. In the 1650s, this ground was protected and raised to the same level as the streets of the chevet of the cathedral. Before the construction of the quays, it was essentially used as a watering place for horses.

The public garden, Square Jean XXIII, next to the chevet and south side of the cathedral, is on the site of several streets which disappeared in the XIXth century. The Palace of the Bishops, then of the Archbishops of Paris was located parallel to the Cathedral in the south, near the Pont au Double. After the burning down of the building in 1831, the archbishopric settled in Rue Barbey de Jouy in the 7th district, where it has remained.

The neo-Gothic building close to the cathedral houses the vestry and the presbytery. It was built in 1845 by the architects Viollet-le-Duc and de Lassus.

The Pont Saint-Louis and the Square de l'Archevêché in 1890. A water-colour by Fraipont.

The Place of Parvis Notre-Dame

It was at the end of the XIXth century that this square was created, bordered in the north by the new Hôtel-Dieu, built under the Second Empire, and in the south by the Préfecture de Police, rebuilt under the Third Republic. In the past this parvis was much smaller than it is today. Except for the towers, the façade of the cathedral was not visible from the quays of the left bank, which was entirely occupied by the buildings of the ancient Hôtel-Dieu.

The parvis such as it is today was refurbished in the 1970s. During the construction of an underground car park, important archaeological discoveries were made. The foundations of the first churches and ancient buildings of the Cité which had themselves preceded the ancient Hôtel-Dieu, were rediscovered and have been conserved.

Notre-Dame of Paris

The construction of Notre-Dame de Paris started in 1163 under Bishop Maurice de Sully. On this site, the Romans had erected a monument to Jupiter, some vestiges of which were rediscovered in the XVIIIth century. Several churches had preceded Notre-Dame on the same site from the IVth to the XIIth centuries. Until his death in 1196, Maurice de Sully supervised the construction which started with the chancel and the transept. The nave and the façade were completed fifty years later, in 1223; the towers in 1250.

The great façades of the transept were built by Jean de Chelles and Pierre de Montreuil, master builders. The apse, with the outstanding design of its 15-metre buttresses was completed in the 1300s. In the XIVth century, a jube was installed inside but it was later destroyed in the XVIIIth century.

The interior of the cathedral was altered in the XVIIth and XVIIIth centuries by J.H. Mansart, Robert de Cotte and Soufflot. Notre-Dame was, of course, hard hit by revolutionary vandalism. The Jacobins of the Convention transformed it into a temple dedicated to the Cult of Reason. At the time of Louis-Philippe and Napoleon III, an opera singer, Miss Aubry, personified Reason.

Victor Hugo gave the most lyrical description of Notre-Dame in his novel «Notre-Dame-de-Paris». This book drew the attention of the public to the very dilapidated state of the building. The resulting stir led to restoration works by the architects de Lassus and Viollet-le-Duc.

Notre-Dame de Paris is now one of the most visited buildings in the capital. The public is attracted there by the prestige and the history of the monument, the homogeneity of the brilliant Gothic architecture, the exceptional set of stained glass windows especially those of the transept, the carved steles of the chancel, the statues and mausoleums by Girardon, Couston and Coysevox; the magnificent great organ restored and reinaugurated in 1992, and, of course, the multitude of statues on the three portals of the façade: Portal to Sainte Anne on the right, Portal of the Last Judgement in the centre, Portal to the Virgin on the left...

The masonry spire built by Viollet-le-Duc is 150 metres high. The nave and the chancel are 130 metres long. The vaults are 33 metres high. The transept is 48 metres wide, the façade and its towers measure 41 metres by 90 metres.

For the most spirited visitors, access to the upper platform of the right tower remains possible via a spiral staircase. From here, there is a splendid view of Paris. These towers are 90 metres high [1].

---

1. Notre-Dame can be visited every day from 8am to 7pm; except during services and concerts. Access to the left tower from the Rue du Cloître-Notre-Dame everyday from 10am to 4.30pm; (Tel: 43 29 50 40).

The crypt of thc Parvis can be visited everyday from 10am to 4.30pm ; (Tel: 43 29 83 51).

The Treasure can be seen on weekdays from 9.30am to 6pm ; (Tel: 43 26 07 39).

Notre-Dame and the Place du Parvis, circa 1895. A water-colour by René Leverdt.

The pont d'Arcole

In 1828 the Seguin brothers built a suspension bridge intended for pedestrians, called the «Passerelle de la Grève» and later, in 1830, the «Passerelle d'Arcole», a man called Arcole having been killed there during the insurrectional days of July 1830. A pier in the river was used as a foundation for a portico on which the chains supporting the deck were fixed. On each side of the pier, the arches had a clear span of 41 metres.

In 1854 this foot-bridge was replaced by a cast iron, single arch foot-bridge with a clear span of 80 metres. It was the first bridge in Paris to be built without supports in the river.

The arch is composed of twelve very thin iron arches. The superstructure, also made of iron, is supported by steel tympana resting on the arches. The handrails are made of moulded cast iron. The bridge is 20 metres wide. Under the supervision of the engineer Oudry, the Pont d'Arcole, begun in 1854, was completed in 1856.

It served for a long time as a night shelter for destitutes who built shelters in the frameworks on each side.

The Pont d'Arcole, the Pont Notre-Dame and the Pont au Change.

The pont Notre-Dame

The pulling down of the houses of the Pont Notre-Dame in 1787.
A painting by Hubert Robert, Musée Carnavalet.

106 metres long and 20 metres wide, the present Pont Notre-Dame was built at the beginning of the Second Empire. This work is on the site of the old Roman «Grand-Pont» which, with the «Petit-Pont», linked the two banks of the Seine.

The Grand-Pont was burnt down by the Normans during the Siege of Paris in 887. It was then replaced by a new work situated lower downstream, on the site of the present Pont au Change. On the site of the first Roman «Grand-Pont» a wooden foot-bridge was built which gave access to a number of floating water-mills. This foot-bridge, washed away in 1406, was replaced by a wooden bridge, the first pile of which was driven in 1421 by Charles VI. Sixty houses were built on it, but they sank into the waters in 1499.

In 1512, the first stone Pont Notre-Dame was built by the Dominican architect Jean Joconde. It was the most harmonious and elegant bridge in Paris with its sixty-eight houses. They had even numbers on one side and odd numbers on the other, a new idea at the time. The two central arches were occupied by the Pump Notre-Dame which disappeared in 1852. The houses had been pulled down in 1787.

The Pont Notre-Dame consists of five masonry arches with a clear span of 18 metres. The semicircular stone pierheads are embossed. Most of the materials of the old bridge were reused in the masonry. The bank arches are embossed; the parapets are in solid masonry.

Reconstruction of the bridge was carried out by the engineers Darcel and de Lagallisserie.

"Floating" and "Hanging" Water-Mills

In the XIXth century large steam-powered mills replaced the mills powered by the stream, which had been so numerous in the past.

From the Middle Ages to the beginning of the XVIIth century, the «hanging» water-mills were the most numerous, replaced gradually by the «floating» mills, both forms thus coexisting for a long time.

The «hanging» mills were built over the river, on wooden stilts or stone piers. In the XVIth and XVIIth centuries they formed actual «millers' bridges». The «floating» mills were placed on the current. They did not create fixed obstacles to navigation and tried to find the best positions.

The most ancient «floating» mills had only one wheel, they were fixed by ropes and chains to piles driven near the banks; access was possible with a «plank» or light gangway. The so-called «double wheel» floating mill was better balanced with its twin paddle wheels and its use became widespread in the XVIIth century.

The «hanging» water-mills whose wheels could be raised or lowered between the stilts on a rack system, according to the height of the river, became common as early as the XIth century.

In order to supply them, the Order of the Temple, was the first to install corn reserves near the Place de Grève in 1147. In 1215 this barn, which belonged to the Temple, adopted the name of «granary on the water» (a street still bears its name to remind us of its existence).

Hanging water-mills had been built very early on the Grand Pont built by King Charles le Chauve (Charles the Bald) (823-877) between the Palace of the Cité and the right bank (Grand Châtelet).

In 1323, to take advantage of the acceleration of the flow, the Prévôt of Paris had a veritable «Millers' bridge» built downstream of the Grand Pont. A series of hanging mills stretched from one bank to the other with the exception of two arches reserved for navigation. A footbridge linked the mills and allowed pedestrians to pass. Being very narrow, it became the first «one-way street» in Paris, the Prévôt des Marchands having stipulated that entry would be on the right bank and exit on the Cité side.

At the end of the XVIth century, the mills not only housed millers but also shops and lodgings. On 23rd December 1596, the «Millers' bridge» collapsed, drowning many inhabitants and passers-by. Little by little the hanging mills disappeared from the Paris scenery to be replaced by floating mills.

However the principle of using water-wheels, for the water pumps, continued up to the XIXth century on the Pont Neuf and the Pont Notre-Dame.

The Millers' Bridge and its hanging mills.

A series of floating mills downstream of the Pont au Change. (optically inversed view)

Water pumps and fire pumps

The Notre-Dame pump in the XVIIIth century. The painting represents a water-joust. Under the second arch, a floating mill. A painting by Raguenet, musée Carnavalet.

The water pump of the Pont Notre-Dame was, with that of the Samaritaine on the Pont Neuf, one of the water raising machines enabling the supply of part of the town with running water.

The Notre-Dame pump was worked by two paddle-wheels which raised the water up to a tank situated on top of a central tower from which a network of tubes and pipes ran.

As far back as 1671, the eight barrels of the pump provided various buildings and fifteen public fountains with 1,600m3 of water a day.

The older «Samaritaine» pump was set up downstream of the second arch of the Pont Neuf on the right-hand bank on the initiative of Henri IV. The machine, which began working in 1608, was designed by a Flemish specialist Jean Lintlaër. The façade, topped by a bell-tower which contained a carillon, was decorated with a high relief representing Jesus and the Samaritan woman at Jacob's well.

The water of the Samaritaine pump supplied the palaces of the Louvre and the Tuileries, but also various public fountains. The station stopped working in 1813.

At the end of the XVIIIth century, the waters of Belleville, the Pré Saint-Gervais, of Rungis, and those of the Notre-Dame and Samaritaine pumps were no longer sufficient to adequately supply Paris. In 1777, two engineers from Grasse founded «La Compagnie des Eaux de Paris» (The Parisian Water Company) with the help of Beaumarchais who looked after the company's «public relations». The Perrier brothers were allowed to build a «fire pump» at the edge of the village of Chaillot, enabling the raising of the water of the Seine up to tanks situated on the hill (the present Place des Etats-Unis).

These pumps worked up to 1853 and were then replaced by new machines which raised 9,500m3 of water a day. The Alma pump disappeared with the preparation of the 1900 World Exhibition.

In 1786, the Perrier brothers' company installed another «fire pump» on the Quai d'Orsay, near the village of Gros Caillou. It supplied the village, the Invalides, the Ecole Militaire (The French Military Academy) and the Faubourg Saint-Germain. The stations worked until 1866.

There was a third fire pump on the Quai d'Austerlitz. As early as 1788 it supplied the Hôpital de la Salpêtrière and the Faubourg Saint-Marcel.

The Samaritaine Pump in the middle of the XVIIIth century. A painting by Raguenet, musée Carnavalet.

The quai aux Fleurs

Begins at the Saint Louis foot-bridge and ends at the Pont d'Arcole.

The Hôtel des Ursins, Quai aux Fleurs.

The Quai aux Fleurs used to stretch right up to the Pont au Change (Money Changers' Bridge), the second part became the Quai de la Corse in 1929. This quay is situated on the site of the former Port Saint Landry which was the Cité's first port. In 1141, Louis VII le Jeune and the Bourgeois of Paris finished it by adding the Port de la Grève. The old Rue des Ursins below the modern quay shows the level of the Ile de la Cité in the Middle Ages.

At n° 9, a plaque and two medallions representing Héloïse and Abélard serve as a reminder that their house stood on this site. At the end of the

The flower-market in 1895. A painting by Firmin Girard.

Middle Ages the biggest construction on the quay was the Hôtel des Ursins which overlooked the Seine and blocked the bank. The first owner of this house was Jean Jouvenel des Ursins, Prévôt des Marchands under Charles VI.

The small Rue de la Colombe which starts at n°21 Quai des Fleurs already had this name in 1223. At n° 4, there is the beautiful old railing from a wine merchant's store. In the middle of the street, different-coloured paving stones show that the wall of the Ile de la Cité passed there.

Behind the quay, at the end of the Rue des Ursins, a beautiful house dating from the beginning of the Renaissance, restored in the sixties by the architect Fernand Pouillon, has become the Parisian residence of Prince Karim Aga Khan. It is called the Hôtel des Ursins.

The ruelle des Ursins.

The flower-market is open every day, from 8am to 7pm, except on Sundays. The bird-market is open on Sundays from 9am to 7pm.

The quai de la Corse

Begins at the Pont d'Arcole
and ends at the Pont au Change.

This quay was built in 1769 and named Quai de la Corse in 1929. It is bordered by the buildings of the Hôtel-Dieu, the flower-market and the Tribunal de Commerce (the Commercial Court).

The stores of the Belle Jardinière, whose name was inspired by the flower-market, were opened in 1824, on the site of the present Hôtel-Dieu. When the Hôtel-Dieu was built, the shop moved to the Rue du Pont-Neuf.

The square Louis-Lépine, home of the flower-market, replaced, under the Second Empire, a maze of particularly unhealthy lanes that also extended along the perimeter of the Préfecture de Police (Police Headquarters), the Hôtel-Dieu, the Tribunal de Commerce, the Rue de la Cité and a part of the square of the Parvis Notre-Dame.

The old quay on the left bank of the Ile de la Cité in the Middle Ages.
Reconstitution by Hoffbauer.

The Dome of the Tribunal de Commerce.

The Tribunal of Commerce

The building of this monument was undertaken by Haussmann who wanted the dome to be an architectural landmark in the centre line of the Boulevard Sébastopol. The architect Nicolas Bailly gave the façade a monumental character with the presence of a sally with columns bearing four stone statues. These statues symbolize «Determination» by Louis Eudes, «Law» by Elias Robert, «Justice» by Jacques Chevalier and «Prudence» by Jules Samson.
The sculptor Carrier-Belleuse achieved an exceptional set of caryatids and allegorical figures placed around the stately central staircase topped by a cupola. A large, well-proportioned patio forms the waiting-hall of the Tribunal. A deep peripheral passage divides the space into two levels, well-decorated with columns. The courtrooms have kept their Second Empire decoration.

The Quai de la Corse in 1885, before the building of the Tribunal de Commerce.

The quai de Gesvres

Begins at the Pont d'Arcole
and ends at the Pont au Change.

This quay was built by Potier de Gesvres under the reign of Louis XIII. It was a long vaulted gallery which covered the low bank. The wall of the present quay was built under the Second Empire at the time of the great works of the «crossing of Paris» round the Place du Châtelet. In 1913, it was decided that the n°7 line of the underground railway wound pass through this tunnel. Since 1860, the buildings lining this quay have replaced the jumble of lanes and the slums of the quarter of the Arcis which stretched from the Hôtel de Ville to the Châtelet.

The Grand Châtelet was destroyed between 1792 and 1802 making way for a square on which a fountain was built. A work by the sculptor Simon Boizot, it commemorated the victories of Napoléon I. It is the «Palm Fountain», so called because of the shape of the column carrying a Victory in gilted bronze.

The vaults of the Quai de Gesvres under which the line n° 7 of the underground railway now passes.

The Grand Châtelet

The fortress of the Châtelet which, between the XIIth and XVIIIth centuries, controlled the north access to the «Grand Pont» or the Pont au Change, was built to defend the Ile de la Cité. Built by Louis VI in 1130, it became, in 1190, under Philippe le Bel, the head office of the jurisdiction of the Prévôté de Paris. Extended, redesigned and restored several times during that century, it was given the name of «Grand Châtelet» in contrast to the «Petit Châtelet» which protected the Petit Pont on the left bank. The main façade faced the end of the Rue Saint-Denis, as shown on the map fixed on the wall of the Chambre des Notaires. A vaulted lane enabled access to the Pont au Change.

The fortress of the Grand Châtelet, end of the XVIIIth century.

Between 1860 and 1862, the architect Gabriel Davioud built the two great theatres bordering the square on the east and west, as well as the Hôtel of the Chambre des Notaires on the north side. The Châtelet theatre with its 1800 seats was the largest theatre in Paris before the completion of the Opera.

The Théâtre des Nations, later the Théâtre Sarah Bernhardt and today the Théâtre de la Ville was completely redesigned in 1966-68, only the façades being retained.

In the centre line of the Boulevard Sébastopol which starts from this square, you can see the pediment of the great hall of the Gare de l'Est built in the same period. On the other side, on the Ile de la Cité, the perspective ends with the large dome of the Tribunal de Commerce.

The Théatre du Châtelet and the Tour Saint-Jacques.

The pont au Change

103 metres long and 30 metres wide, the Pont au Change has three 32-metre elliptical arches. It was rebuilt by the engineers Romany and Vaudrey in 1858-60, at the time of Baron Haussmann's extensive replanning of Paris, in order to be in line with the new Boulevard du Palais and in the centre line of the new square of the Châtelet. It replaced the bridge which Jacques-Androuet du Cerceau had built in the middle of the XVIIth century.

The first Pont au Change was built in about the year 1000. Carried away by the river in 1280, it was rebuilt in 1296 and was then called the «Pont du Roy». From the XIVth to the XVIIth centuries, it was inhabited by goldsmiths and money-changers, from whom it takes its name.

Swept away in 1616, it was rebuilt by the architect Jacques-Androuet du Cerceau. It was a large stone bridge with houses on each side.

The architecture of the present work, like that of the Pont Saint-Michel, is characterized by its stone balusters, bossed outer arches and tympana decorated with the monogram of Napoleon III engraved in a laurel wreath, sculpted by Cabat.

The Pont au Change and its houses in the XVIIth century. On the left, the Notre-Dame pump.
A painting by Schauwaetz, Musée du Mans.

The Pont au Change, the Tribunal de Commerce and the Clock Tower, circa 1890.

The Pont au Change after its reconstruction during the Second Empire.

The pont Saint-Michel

From 1364 to 1617, several wooden bridges were built successively on the same site ending with a four-arched masonry bridge. 25 metres wide, it bore two rows of thirty-two houses which only disappeared in 1807.

Downstream, on the central tympanum, there was an equestrian statue of Louis XIII, destroyed during the Revolution.

At the time of the designing of the Place Saint-Michel, the bridge was found to be wrongly aligned and on the wrong level. It was therefore rebuilt in 1857.

It consists of three 17-metre elliptical arches with two 3-metre thick piers. The thickness at the keystone is only 70 centimetres. The structure is 30 metres wide between the parapets. These, with their balusters are, like the tympanums, the bossed outer arches and the semicircular piers, in ashlar.

Above the semicircular sallies covered with bosses, laurel wreathes are sculpted around the letter N for Napoleon. The parapets with their balusters are made of stone.

The sculptures are by Lavigne. The very rapid construction, led by the engineers Romany and Vaudrey, only lasted five months, from May to September 1857.

The Pont Saint-Michel, circa 1840. A painting by Masson, Musée Carnavalet.

The Pont Saint-Michel and the Quai des Orfèvres, circa 1912. A water-colour by Leverd.

The quai du Marché-Neuf

Begins at the Petit Pont
and ends at the Pont Saint-Michel

The long façade of the Préfecture de Police takes up almost the whole length of the quay. In the years 1870-80, after the fires of the Commune, it was rebuilt in a neo-Classical style around a vast main courtyard. The monumental gateway faces the Place du Parvis. Up to 1808, houses were built along the river, often overhanging it. Construction of the present quay began during the First Empire and was finished in the 1880s.

Théophraste Renaudot founded the first newspaper in the world, La Gazette de France, in a house which stood on the site of the present n°8. Doctor and philanthropist, he managed this newspaper until his death in 1653. Publication continued for a century after his death.

The Quai du Marché-Neuf, built on land reclaimed from the river, has only a very narrow bank which stretches to the Pont au Double below the Parvis Notre-Dame.

The Quai du Marché-Neuf and the Préfecture de Police in 1890. A drawing by Fraipont.

The Hôtel-Dieu

The old Hôtel-Dieu of Paris was built on the small arm of the Seine between the Place du Parvis Notre Dame and the Rue de la Bûcherie, from the Pont au Double to the Petit Pont.
Its origin goes back to the beginning of the Middle Ages. It was rebuilt in 1165 by Maurice de Sully at the same time as Notre Dame, then restored and extended at the beginning of the XVIIth century. At the beginning of the XIXth century this hospital had 2,200 beds. At that time, it was one of the biggest. It was pulled down in 1878 after the construction of the new hospital which stands on the other side of the Place du Parvis.
It is quite an outstanding construction which has often been used as a model. Haussmann insisted that the buildings should be quite low, only two storeys, so as not to block the sunlight from the huge peripheral courtyards. The principal courtyard is surrounded b wide covered walk, whose columns and archways give a most harmonious effect.

The quai des Orfèvres

Begins at the Pont Saint Michel
and ends at the Pont Neuf.

The collection of buildings which makes up the Law Courts constitutes one of the biggest and most complex groups of monumental buildings in Paris. It was built at different periods from the XIIth century to the present time. Before summing up its history, let us situate it within the context of the quays that surround it.

Photographs and paintings from the last century show that the Law Courts' buildings have been substantially altered. Between the Sainte-Chapelle and the Seine, the successive extensions were only completed before the First World War.

For centuries, houses stood directly above the bank of the small arm on the upstream part of the quay, along the now inexistant Rue Saint-Louis.

The money changers and the goldsmiths settled in this street when the Pont au Change was carried away by the flood in 1616.

The Sainte-Chapelle

Built on the initiative of King Louis IX, Saint Louis, the Sainte Chapelle is divided onto two levels: the lower level, initially reserved for the staff of the Palais; the upper level for the King and his Court.

It was built to shelter Christ's Crown of Thorns and the Passion relics that Saint Louis had acquired in 1239 from Beaudoin II, Emperor of Jerusalem.

This masterpiece was attributed to the architect Pierre de Montreuil who also worked on Notre Dame. The works lasted only five years, from 1243 to 1248.

The Sainte Chapelle narrowly escaped demolition during the Revolution. The spire was pulled down; the treasure plundered

The Quai des Orfèvres in 1850. A painting by Masson, Musée Carnavalet.

The Pont Saint-Michel, the Sainte-Chapelle. A drawing by Fraipont.

The presence of these goldsmiths was the reason for the naming of the quay, whose construction began around 1620, when the houses of the Place Dauphine and of the Rue de Harlay were being built.

The building of the houses on the Place Dauphine, named in honour of the Dauphin Louis XIII, the son of Henri IV, was carried out by the master mason François Petit, under the supervision of President Achille de Harley, entrusted by Henri IV with this prestigious town planning operation.

At the corner of the quay and the Boulevard du Palais stands the «Tribunal Correctionnel» built by Albert Tournaire between 1907 and 1914. A certain number of sculptures enliven its austere façade. Four very graceful female figures of the 1900 style represent «Truth» by Henri Lombard, «Law» by André Allard, «Eloquence» by Raoul Verlet, «Mercy» by Jules Coutant. A sundial was decorated with a high relief by Antoine Injabert: «Time and Justice». Above the main gateway at

and scattered. The restoration, undertaken by Louis-Philippe, was given to the architects Duban and de Lassus who restored the spire and repaired the XIIIth century stained glass windows in which the master glassmakers of the Middle Ages had represented, in 1,134 scenes, the relationship between the Old and New Testaments. These are the oldest stained glass windows in Paris. The fifteen large stained glass windows cover some six hundred square metres.

Little by little, closed in by the buildings of the Law Courts, only the 75-metre high spire of the Sainte Chapelle is visible from the outside.

The Sainte Chapelle can be visited every day from 10am to 4.30pm. Entrance : n° 4 Boulevard du Palais. Tel: 43 54 30 09.

The Quai des Orfèvres and the Tour Pointue in 1980.

A Sunday, Square du Vert Galant in 1907. A painting by Ferdinand Frambusch.

the corner of the Boulevard du Palais, there is an allegory.

These buildings are topped by a high square tower with a very pointed roof, a sort of view-point over the Seine and the Law Courts. On the downstream side, towards the Pont Neuf, all the houses of the beginning of the XVIIth century have been raised. Only the two houses facing the Pont Neuf and the statue of Henri IV have kept their original proportions. The same has happened on the Quai de l'Horloge. The row of buildings which formed the third side of the Place Dauphine was pulled down in 1872 to allow the widening of the Rue de Harlay, after the construction of the buildings of the Supreme Court of Appeal.

The flower-bed of the Pont Neuf as seen from the Quai de Conti in 1900. A water-colour by Leverd.

Les bouquinistes (Second-hand booksellers)

Anatole France, who, from his windows of the Quai Malaquais, could see them every day regardless of the weather, always sitting in front of their open boxes, used to call them «Merchants of wit» ; Apollinaire used to speak of the charming public library that could be consulted every day. «I would wander along the bank of the Seine. An ancient book under my arm.» Francis Carco, Roland Dorgelès and Léon-Paul Fargue, were, in the past, regular clients of the second-hand booksellers as was Alexandre Arnoux who in one of his best books «Paris, ma Grande Ville», conjures them up in saying: «The genuine quays, according to my own heart, are only on the left bank from the Pont Royal to the Pont aux Doubles, with real second-hand booksellers and loyal and true book-hunters. The right bank has only second-hand bric-a-brac dealers, used books and hoarders of old papers. No, let us not leave the left bank. There, among the venerable muddle and the treasures, the diamonds of typography shy away from the uninitiated...

... If book-hunters are recruited among all the social classes, where do the second-hand booksellers, an homogeneous race of different individuals but of the same innate character come from? Most of them are supposed to come from Normandy...

The tradition of second-hand booksellers in Paris is very old. The first ones might have been itinerant merchants, street pedlars offering second-hand books from wicker-baskets hung round their necks. A ruling in 1578, at the beginning of the reign of Henri III, forced them to settle; but it was only at the beginning of the XVIIth century, in 1619, that the second-hand booksellers settled on the quays, and more exactly on the Pont Neuf, where all the small semicircles that rise above the piers offered convenient recesses. Little by little, boxes took over the quays, beginning with the left bank. Balzac, in the 1840s, evoked the lovers of rare books, milling around the bookstalls near the Pont Royal («La Peau de Chagrin»).

Bouquinistes at the corner of the Pont Neuf and the Quai de la Mégisserie.
A painting by Brispot, circa 1900.

At the time of Balzac and Stendhal, the second-hand booksellers were also present under the galleries of the Palais Royal which was, at that time, one of the centres of intellectual and social life, and where the publishers could be found. They could also be found under the arcades of the theatre of the Odeon, from where they only disappeared at the beginning of the XXth century.

Today the second-hand booksellers' boxes are still more abundant on the left bank. They can be found continously from the Pont Sully to the Pont Royal, whereas on the right bank, they can only be seen at the Pont Marie, not quite reaching the Pont des Arts. In 1993, there were about two hundred and fifty second-hand booksellers in Paris.

Strangely enough, there are no bookstalls on the quays of the Ile Saint-Louis and of the Ile de la Cité, where the second-hand booksellers and clients would be better protected from the bustle and the dangers of the traffic...

Bouquinistes on the Quai de la Mégisserie, circa 1900. A painting by Galien Laloue.

Bouquinistes on the Quai Montebello, circa 1910. A painting by Edouard Cortés.

The quai de l'Horloge

Begins at the Pont au Change
and ends at the Pont Neuf

Allegory of Justice on the Quai des Orfèvres.

The Quays of the Cité. Reconstitution by Hoffbauer.

This quay was built between 1580 and 1610 with significant alterations in the XIXth century, at the time of the reconstructions of the Palais and the restoration of the Conciergerie. It was called «Quai du Grand Cours d'Eau» because it was on the side of the main arm of the river and later the «Quai des Morfondus». Its present name is due to the presence of the monumental clock near the Pont au Change.

The Clock tower, as well as the marvellous kitchens of the Conciergerie, were built by Jean Le Bon in the middle of the XIVth century.

This was the first public clock in Paris. Commissioned by Charles V in 1371, it is still working.

The prison of the Conciergerie originated from the XIVth century at the time when the «concierge» (keeper) of the Palais, generally a person of high standing, had a right of justice which involved the presence of a prison for criminals arrested in the Palais and its vicinity. The prison of the

On the preceding page : Panoramic view over the Ile de la Cité, from the right bank, at the three different periods.

The Clock Tower and the Dome of the Tribunal du Commerce.

Conciergerie rapidly became a prison annex for the Grand Châtelet, the prison of the Parisian Parliament.

During the Reign of Terror, the revolutionary Court was installed in the Palais; presided over by Fouquier-Tinville, it pronounced death sentences on 2,278 persons including Queen Marie-Antoinette.

The best known and most visited part of the Conciergerie is the magnificent Hall of the «Men at Arms» which is sixty-nine metres long, twenty-seven metres wide and eight metres high. Three rows of eight pillars divide it into four naves with diagonally-ribbed vaults. Built between 1301 and 1312, it was originally used as a refectory for the staff of the Palais. Nowadays it is used for concerts and temporary exhibitions. The ground level, much lower than that of the Boulevard du Palais, shows the average level of the ground of the Ile de la Cité in the days of Philippe Le Bel.

The Quai de l'Horloge and the Pont Neuf as seen from the Quai de la Mégisserie, circa 1840. In the foreground a floating wash-house. A painting by Giuseppe Canella, Carnavalet.

The Gothic-style «Bombec» Tower is the oldest in the Palais. Built under Saint Louis, it was integrated into the constructions undertaken by Philippe Le Bel from 1285 to 1314, the Hall of the Men at Arms, the Tour de César (the Caesar Tower) and the Tour d'Argent (the White Tour). Restored in 1828, it was thus named because it was used as a «torture chamber», the accused having «bon bec», a knack of speaking under the pressure of the various tortures which were commponplace in the Middle Ages.

West facade of the Palais de Justice, Cour de Cassation.

The kitchens, like the clock towers, date back to King Jean Le Bon (middle of the XIVth century). They form a square with 17-metre long sides. Three rows of columns divide the space into four naves with diagonally-ribbed vaults. At the corners are four huge chimneys whose hoods are supported by stone buttresses.

The façades between the four towers on the quay upto the Rue de Harlay were restored in the XIXth century; a neo-Gothic style for the Conciergerie and an eclectic neo-Classical style for the Supreme Court of Appeal. The latter is distinguishable by the slightly prominent central pavilion decorated with allegorical statues by Eugène Lesquesne. They represent «France», «Justice», «Innocence» and «Crime». The pediment is an allegory of «Law protecting Innocence and punishing the Guilty».

The Palace of Justice (the Law Court) is extended by its façade to the Rue de Harlay where the architect Duc built a colonnade which faces the charming Place Dauphine. Between the columns, six allegorical statues symbolize «Prudence» and «Truth», by Auguste Dumont, «Chastisement» and «Protection» by François Jouffroy, «Strength» and «Equality» by Jean-Louis Jaley. The two big stone lions lying at the foot of the grand staircase are by Isidore Bonheur.

Access to the Law Courts is principally on the Boulevard du Palais side, through the main courtyard, which is also called «Cour du Mai» (the May Courtyard).

The gate of the Law Courts, the May Courtyard.

In this guide we can only rapidly describe the architecture and the complicated history of this building which was the first Parisian palace of the kings of France.

Historians generally trace the presence of an official residence back to Gallo-Roman times - Julien, governor of the Gauls, lived there in 355. In the Vth century, Childebert I, Clotaire I, Chilpéric I, Frédégonde and Clotaire II lived there and in the VIIth century Dagobert I and his son Clovis II. In the VIIIth and IXth centuries, Pépin Le Bref, Charlemagne, Louis Le Pieux and Charles Le Chauve occasionally lived there.

Towards the year 1000, Robert II, son of Hugues Capet had the «Kings' Hall» and the «King's palace» built with gardens on the west side. Philippe Auguste (1165-1223) lived in the palace and had the neighbouring streets paved. He also tried to introduce streetlighting.

Louis IX (Saint Louis, 1226-1270) built the Sainte Chapelle and the «Galerie des Merciers» (Drapers' Gallery) at the end of the May Courtyard. (Every year in May a tree symbolizing Spring and the Revival was planted).

His grandson Philippe Le Bel (1268-1314), found the palace too small and, in 1298, commissioned Enguerrand de Marigny, his principal minister (1260-1315) to have it reconstructed in order to place the government, the judicial and financial services within the Royal Residence. It was at that time that they built the Caesar Tower, the White Tower, the Guardroom and the Hall of the Men at Arms with the Great Hall just above on the first floor. This was later replaced by the Lobby. At the beginning of the XIVth century Jean Le Bon built, amongst other things, the kitchens and the Clock Tower.

Charles VII (1403-1461) left the palace to live at the Louvre, leaving it definitely to Parliament in 1431. The High Hall or the «Great Hall» of Philippe Le Bel was destroyed by fire in 1618. The stone statues of fifty-eight kings of France and the «Table de Marbre» where official meals were served at the time of the kings and where

The May courtyard

The splendid wrought iron railing that separates the Courtyard from the boulevard was made in 1776 by Bigonnet following the designs of the architect Desmaisons. At the back of the yard, the majestic staircase leads to the gallery of the Palais, whose façade is distinguishable by its projecting part with four Doric columns. On the cornice, statues represent «Abundance», «Prudence», «Justice», and «Strength».

The Quai de l'Horloge and the bridges on the right bank arm.

The Quai de l'Horloge, as seen from the Samaritaine.

jokes and tricks were played, disappeared at the same time. The Drapers' Gallery with its numerous shops connected the Great Hall with the Sainte-Chapelle.

The architect Salomon de Brosse rebuilt the Great Hall in 1622. Burnt down by the Commune in May 1871, it had to be rebuilt by the architects Duc and Daumet in 1886-88. It is now the present «Salle des Pas Perdus» (lobby).

The towers and the pepper box turrets of the Conciergerie.

The conciergerie can be visited daily from 10am to 5pm. The entrance is at n°1, Quai de l'Horloge. Tel: 43 54 30 06.

The quai de la Mégisserie

Begins at the Pont au Change
and ends at the Pont Neuf

The Châtelet and Saint-Jacques Tower.

This quay, built for the first time in 1369, was one of the oldest in Paris with those of the Cité. It was completely rebuilt under the Second Empire. It gets its name from the presence of «mégissiers» (leather-tawers) who occupied the banks of the Seine on this site for more than five hundred years. At the beginning of the XIXth century, they settled on the banks of the Bièvre, near the Gobelins.

This quay was also called «Quai de la Ferraille» owing to the presence of second-hand iron dealers.

The high neo-Classical buildings decorated with Corinthian pilasters which overlook this quay, were built under the Second Empire.

Panorama of the quai de la Mégisserie.

Le pont Neuf et le grand magasin de la Samaritaine.

Along the banks, as on most of the south-facing banks of the right bank, there were a lot of floating wash-houses, the last of which only disappeared in the thirties.

The parapets were, very early on, occupied by the second-hand booksellers whereas the bird-sellers kept their shops at the foot of the buildings.

This tradition still continues today with the presence of corn chandlers and nursery gardeners who continuously display nany varieties of plants, ornamental shrubs and flowers along the pavements. Some businesses, like Vilmorin and Andrieux have been settled here for over a century.

The **«Grand Pont»** built in the middle of the IXth century by Charles Le Chauve was rather low so that whenever the river was very high, its numerous piers formed a barrier and the flooded deck became impassable. This lasted for more than two centuries. In 1142, King Charles VII built a sort of wooden upper bridge, whose deck bore houses and shops, where Jewish and Lombard money-changers settled. The old «Grand Pont» was then called the «Pont des Changeurs»and was the first bridge in Paris to bear houses. The Pont aux Changeurs was carried away by a flood in 1510 and re-established fifty-six years later in 1566, then restored in 1616 before disappearing in a fire in 1621. Rebuilt during the reign of Louis XIII between 1639 and 1647, it was called the «Pont au Change» although the money-makers were not the only people present there. It carried two rows of tall, narrow houses which like those of the Pont Notre-Dame, the Petit Pont, the Pont Saint-Michel and the Pont Marie, disappeared just before the Revolution.
The Pont au Change was rebuilt by Haussman in 1858 in its present form.

The quai des Grands-Augustins

Begins at the Pont Saint Michel
and ends at the Pont Neuf.

The Lapérouse Restaurant

The chronicle of the establishment states that the beautiful women who used to have lunch or dinner at Lapérouse engraved their initials on the mirrors of the salons with the diamonds offered by their admirers, in order to check their authenticity...
At the end of the XIXth century, this habit still existed but at that time it was Dumas, Maupassant, Zola, Victor Hugo and so many others from the litterary and artistic worlds who visited the Lapérouse salons. The panelling and the ceilings were restored at that time and decorated with light, rustic scenes in the style of

This quay has been so called since the XVIIth century. It offers a succession of high houses of the XVIth, XVIIth, XVIIIth and XIXth centuries which follow the bend of the Seine and form a beautiful panorama. It is one of the oldest roads in Paris, built for the first time in 1313 by Philippe Le Bel to link the Palais de la Cité and the Hôtel de Nesle, by passing over the Petit Pont, the only one that existed at that time on the left bank.

At n°1 is a restaurant «La Rôtisserie Périgourdine»; at n°15 is the Cabaret de l'Ecluse where many contemporary singers began, from Jacques Brel to Barbara, from Georges Brassens to the Frères Jacques.

N° 35, at the corner of the Rue Séguier is a beautiful mansion from the end of the XVIIth century. In 1740, the famous printer and type founder, François Didot moved in there. At n°41, a big

The Quai des Grands Augustins in 1880. A painting by Jules Trayer.

The Quai des Grands Augustins. A moored tug. A painting by Madelain, circa 1900.

Louis XIV style house is occupied by the La Pérouse restaurant. The windows railings of the windows and the sign pay homage to the famous explorer Jean François de Galaup, Comte de la Pérouse (1741-1788). N°51 boasts two beautiful stone caryatids on either side of the portal, on the entresol.

Between the present n°53 and the Rue Dauphine, on a perimeter stretching south to the Rue Christine, there was, before the destructions of the Revolution, one of the most beautiful churches in Paris and one of the most important convents in France, Les Grands Augustins, founded in the XIVth century. Some relics of the abbey, columns, vaults and capitals, remain in the cellars at n°53b and 55-57. It was in this abbey that Marie de Médicis was proclaimed Queen Regent in 1610, after the murder of Henri IV.

The headquarters of the Compagnie du Métropolitain de Paris was built in the 1920s, at n°53a, which later became the building of the RATP management.

The buildings that border the Rue Dauphine were built in the 1930s by the architect Joseph Marrast. In the left-hand building is the Pont Neuf bookshop, one of the best bookshops for old books. Under the one on the right is the entrance of the strange, very narrow old Ruelle de Nevers. In the XVIIth century, it delimited the Hôtel de Nesle, later called the Hôtel de Nevers.

Watteau. Registered in the Inventory of Historic Monuments, these salons have remained among the high circles of Parisian gastronomy where, for ages, people have seen the history of literary, political and society life in Paris.

At the beginning Lapérouse was a cabaret created by Lefebvre, a lemonade manufacturer at the time of Louis XIV, in a mansion, which had belonged to the Comte de Bruillevert, Maître des Eaux et des Forêts (Master of the Waters and Forests). Jules Lapérouse, namesake of the famous sailor, transformed the premises by creating small salons so that the businessmen and authorized sellers of the poultry market, which was on the Quai des Grands Augustins, might conduct their business and also receive their mistresses. A secret staircase enabled people to leave the palace without being seen, when necessary, across the backyards beyond the convent of the Grands Augustins, which was destroyed at the Revolution.

A mansion on the Quai des Augustins.

The pont Neuf

The oldest of the bridges in Paris was begun under the reign of Catherine de Medicis and Henri III, in 1578. The work took thirty years to complete. The Place Dauphine and its houses were built at the same period. The architects of The Pont Neuf, Guillaume Marchand, Thibeau Métézeau, Pierre Chambiges, François des Isles and Androuet du Cerceau were also entrusted with the vast town planning operation which depended on the construction of the bridge. It consisted of creating a new district at the downstream tip of the city and improving access between the Louvre, the Saint-Germain-des-Près Abbey and the left bank.

King Henri IV inaugurated the Pont Neuf in 1607. Wide and convenient, it was the first bridge without houses. Parisians could cross the river while enjoying the view. This bridge had wide stone, not wooden pavements, which was an innovation...

The Pont Neuf and the Ecluse (lock) de la Monnaie in 1910

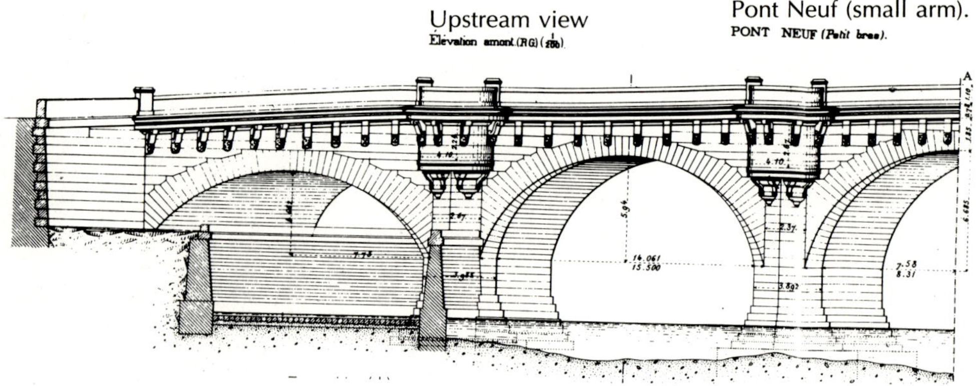

Small trade was prosperous, the second-hand booksellers were the first to set up their businesses as early as 1619. In 1675, twenty booksellers' shops were created in the half-moons surmounting the piers. They remained there until the great restoration of the bridge under the Second Empire.

The construction of the Pont Neuf left the Ilot de la Gourdaine untouched in its downstream part. It became a public garden, the Square du Vert Galant, so named because of the statue

Henri IV on his bronze horse as seen as from the Place Dauphine.

The Pont Neuf as seen from the Quai du Louvre at the end of XIXth century. A painting by Noël, Carnavalet.

of Henri IV in the flowerbed. Its level has not really been changed.

In the XVIIth and XVIIIth centuries, the Pont Neuf was, like other bridges, a site for public exhibitions by the painters of the Saint-Luc Academy who sold their pictures there, out in the open. It was also one of the most privileged places for the setting up of architectural decorations erected for royal coronations, princely weddings and the reception of foreign sovereigns. The Seine, with the majesty of its decorations, offered a large space for all the water shows and firework displays.

A few arches of the Pont Neuf.

Traffic on the Pont Neuf, circa 1890.
A painting by Guillermot.

The Pont Neuf seen from the bell tower of the Samaritaine. A painting by Serge Belloni, 1973.

The work is divided into two distinct parts: the bridge on the left bank with four arches, the bridge on the right bank with seven arches. It is 232 metres long and 22 metres wide. Under the Second Empire, it was provided with new streetlamps designed by the architect Victor Baltard.

The equestrian monument on the flowerbed, representing Henri IV, was sculpted by François Lemot in 1818, at Louis XVIII's request, to replace the statue destroyed during the Revolution.

The former was the first equestrian statue erected in Paris, in 1614, four years after Henri IV's death. It was the work of the Florentine caster Pierre Tacca, from a model by the sculptor François Franqueville, himself a student of Jean Bologne.

The pont des Arts

The Pont des Arts was an avant-garde technical achievement. A link between the right and left banks between the Cour Carrée du Louvre and the Palais de l'Institut, it is 155 metres long and 10 metres wide. Right from the beginning, this work was reserved for pedestrians . A toll of one sou was payable until 1848. Its construction, between 1801 and 1803, was supervised by the engineers Louis Alexandre de Cessart and Jacques Lacroix-Dillon. It was the first iron bridge thrown in France. It is a work of great technical and aesthetic finesse. Situated exactly in the line of the Cour Carrée and that of the Dome of the Institute, this bridge is substantially higher than the quays. Originally it had nine uniform arches each spanning 17 metres. It was rebuilt in 1982-84 under the supervision of the architect Louis Arretche, with one less arch, to make navigation easier.

The Palais du Louvre which became the «Museum National» under the Convention, was also called the «Palais des Arts» under the Empire. The bridge which led to it was naturally called the «Pont des Arts».

The pont des Arts.

The Pont des Arts and the Pont du Carrousel.

The Pont des Arts. Opposite the Institut.
A drawing by Fraipont.

The pont des Arts.

The pont du Carrousel

The ancient Pont du Carrousel, which disappeared in 1936, was a real work of art because of its technical and architectural finesse. It was built at the beginning of Louis Philippe's reign. It was also called the «Pont des Saint-Pères» due to the proximity of the street to which it led.

It had three depressed iron arches spanning 47 metres, each of them consisting of five trusses resting on two 4-metre thick piers and on the

The dock of the Louvre as seen from the Pont du Carrousel in 1910. A water-colour by Leverd.

The ancient Pont des Saint-Pères or of the Carrousel, circa 1890.

The Pont du Carrousel and the «entrance gate» of the Louvre.

abutments of the banks. These trusses were a technological innovation that has not since been repeated. They were, in fact, wooden structures nailed and cemented to the asphalt, consisting of eleven layers of wooden beams faced with bolted cast iron plates; the profiles of these trusses were elliptical. They were only twelve metres wide which resulted in the bridge's pulling down when car traffic increased in 1935. The construction was led by the engineer Antoine Rémi Polonceau.

In 1847, four statues by Léon Petitot were placed at the corners, allegories of «the Seine» and «the Ville de Paris» on the left bank, of «Industry» and «Abundance» on the right bank.

168 metres long and 33 metres wide, the new Pont du Carrousel was built in 1936-1937. It is a concrete work with stone facing. Four monumental telescopic streetlamps designed by the wrought iron sculptor, Raymond Subes, were supposed to light the whole bridge.

The neighbouring Place du Carrousel, behind the entrance gate of the Louvre, was the origin of the bridge's present name, the work having previously been called the «Pont des Saint-Pères».

The quai de Conti

Begins at right angles with the Pont Neuf and ends at the Pont des Arts.

Built from 1662 onwards, it was successively called Quai de Nesle, de Nevers, de Guénégaud, des Quatre Nations, de la Monnaie and finally, since 1815, Quai de Conti.

In the days of Louis XIV and Louis XV, the Hôtel des Monnaies (the Mint) was situated on the right bank near the Pont Neuf. Having fallen into disrepair, it was decided to transfer it to the Quai de Conti. The architect Jacques-Denis Antoine was entrusted with the project. The first foundation stone was laid in 1771. The construction was rapid and was finished in 1777.

This palace is the most significant building of the transitional architectural style of the second third of the XVIIth century which would become the Louis XVI style. The beautiful long façade on the Seine (110 metres) is of a rare elegance. The

The Tower and the Hôtel de Nesle

Jeanne de Bourgogne, famous for her love affairs, notably with a student named Buridan, lived in the Nesle Tower between 1322 and 1329. She was a prisoner there. In 1435, the Duc de Berry, Charles V's brother, had a far bigger residence erected on the site of the first Hôtel de Nesle. Isabeau de Bavière organized feasts there, Charles le Téméraire (the Bold) stayed there in 1461. It also housed, in 1549, Benvenuto Cellini's coin-workshop.

In 1572, Louis de Gonzague, Duc de Nevers, had these buildings pulled down and replaced with sumptuous buildings which were used for the love affairs of Henriette de Clèves and Anne de Gonzague. A century later - in 1676 - Marie Louise de Gonzague had the magnificent Hôtel de Nevers pulled down and the Rue Guénégaud opened.

The Tour de Nesle. Reconstitution by Hoffbauer.

The Pont des Arts and l'Institut, circa 1880. A painting by Delpy.

three levels of windows, the extremely prominent cornice with its consoles give the impression of lengthening the building even more. The vertical lines are repeated without monotony. The presence of the large central sally part balances out the construction.

This sally has six high Ionic columns above the five circular archways on the ground floor. Six monumental allegorical statues form the corona; from left to right they represent: Abundance, Peace, Trade, Strength and Justice. The magnificent composition of the recesses matches the harshness of the façade. The quality of the decoration helps to turn this edifice into a real masterpiece of French architecture and decorative art.

The coffered peristyle of the entrance hall opens on to the semicircular main courtyard, surrounded by a gallery with balusters. The staircase is one of the most imposing in Paris.

The great central room is unique by the layout of its architecture: twenty Corinthian columns support a very prominent entablature on which a gallery is set up. The room is crowned by a painted elliptical cupola.

The Palais de l'Institut.

The Square and the Palais de l'Institut under the Second Empire. A print.

The Academies

The buildings of the ancient college founded by Mazarin, now house the five Academies which form the Institute: the French Academy, the Academy of Fine Arts, the Academies of Inscriptions and Belle Lettres, Science, Moral and Political Sciences. The Mazarine library in the right wing of the buildings, facing the Seine, was created for the collections of Cardinal Mazarin and has been open to the public since 1643. Reinstalled in 1691 in the Palais des Quatre Nations, it was one of the biggest libraries in Europe. Today it holds more than four hundred thousand books.

Condorcet's statue on the Quai de Conti.

The succession of exhibition rooms on the quayside is today occupied by the coin museum, created by Charles X in 1827. This museum displays an exceptional collection of coins and medals explaining the evolution of the art of coin-making from the beginning to today.

In the workshops of the Mint, which occupy a huge perimeter around several courtyards, commemorative medals and certain collection coins are still made.1

Between the Mint and the Institute of France there is a small square set back from the traffic where three buildings of the XVIIth and XVIIIth centuries stand. In 1991, the Mint and the Ville de Paris had a bronze statue erected there. The statue represents the Marquis de Condorcet (1743-1794), a mathematical philosopher and politician. President of the Assemblée Legislative in 1792, he was poisoned and subsequently died at the end of the Reign of Terror.

In the Middle Ages, the Hôtel de Nesle stood on the site of the Institute. The Tour de Nesle was a huge building, with a diameter of 10 metres and a height of 20 metres, from which the Seine could be surveyed. According to the plan fixed on the wall of the building, it was on the site of the left wing of the Institute relative to the bridge.

The Tour de Nesle disappeared in 1663 as did the Philippe Auguste ramparts. The foundations of this fortification are used as foundations for the whole of the Rue Mazarine on the left-hand side as you head upstream. Large parts of it are visible, cleared during the constuction of the underground carpark just under the Passage Dauphine.

A theatre-barge on the Quai de Conti.

The Palais of the Institute of France was originally a project by Cardinal Mazarin who decided in 1661 to create a college of «Four Nations» intended to welcome students coming from the four groups of provinces or «Nations» recently joined into a kingdom. Jean-Baptiste Colbert entrusted Louis Le Vau who was then working on the Louvre, with the construction of the college, which began in 1663.

The College of Four Nations was designed as counterpoint on the left bank to the new buildings of the Louvre. The line of the Cour Carrée corresponds to that of the Institute . The architect François d'Orbey completed the building in 1674. The college was opened in 1688.

In 1805 Napoléon decided to house the Institute of France there. The Chapel was altered by the

The Hôtel des Monnaies (The Mint).

architect Vaudoyer who made it into the Session Room. In the sixties André Malraux had the premises refurbished, restoring le Vau's architecture to its original splendour.

The perfectly symmetrical buildings are laid out in a semicircle around the square which is in front of the central portico. The space encompassed by the symmetrical wings and the square pavilions which end them, creates a deep recess, unique when compared with the usual alignement of the buildings on the quays of Paris. This layout enhances the monumental character of the construction. The large curved wings with their flat rooves decorated with balustrades and large «vases de feu» (vases in which fires were lit to light the area), are crowned by the dome and its lantern. This round dome shelters an elliptical cupola on the inside.

---

1. In the Rue Guénégaud, coins and medals of the past and present are on display and on sale. Open from Monday to Saturday 10am to 1pm and 2pm to 5pm.

The Coin Museum is open every day from 1pm to 6pm except on Mondays. The rooms are also open for temporary exhibitions.

Skating on the dock of the Louvre, January 4th 1880.

The quai Malaquais

Begins at the Pont des Arts
and ends at the Pont du Carrousel.

The Quai Malaquais and the ancient Pont des Saints-Pères in 1886. A water-colour by Harpignies.

The building of this quay goes back to the middle of the XVIth century. At the beginning it stretched as far as the Pont Royal. It was named «Quai de la Reine Marguerite de Valois» in the XVIth century. The mansion of Queen Marguerite de Valois -(1553-1615), the first wife of Henri IV- was situated at the corner of the Rue de Seine. She had the convent of the Petits Augustins built in an annexe in 1608. Only the Chapel remains, situated within the walls of the School of Fine Arts.

The Queen Marguerite park was parcelled out in the middle of the XVIIth century. The Quais Malaquais and Voltaire, the Rues de Lille, de Verneuil and de Beaune can now be found on this site.

The bank of the Quai Malaquais in 1874. A painting by Harpignies.

The Hôtel Dorat, at n°3 Quai Malaquais, is a beautiful construction from the second half of the XVIIth century. The noble floor is enhanced by a beautiful wrought iron balcony. A plaque decorated with a high relief medallion indicates that the explorer Alexandre de Humbolt, a member of the Institute (1769-1859), lived in the house.

Built by Joseph Dorat, a Commissioner of Audit, the townhouse at n°3 was occupied by the Maréchal de Saxe, the grandfather of Georges Sand. The painter Joseph Vien had his studio there.

The Hôtel de Châteauneuf at n°5 is a very elegant building with five high semicircular arches on the level of the ground floor and the entresol. The noble floor is characterized by the particularly skilful design of its windows and the delicacy of their balconies.

Inside, the magnificent main staircase and its wrought iron banisters, the large first floor rooms and their ancient woodwork and decorations match the quality of this noble façade.

The adjacent mansion at n°7 was built in 1622 by Jacques de Garsanlan, a Councillor and Master in Ordinary to the Chambre aux Deniers. On the ground floor there is the café Léon Malaquais frequented by many generations of Fine Arts students. The famous old publishing firm Honoré Champion is also housed in this building.

On the small square situated near the left wing of the Institute, a stone copy of the statue of the Republic has been erected. It was sculpted in 1880

Statue of the Republic on the Quai Malaquais.

The Quai de Malaquais in 1905. A painting by Leverd.

by J.R. Sartoux. This monument was inaugurated by Jacques Chirac, the Mayor of Paris in 1992.

The Hôtel de Transylvanie, situated at n°9, owes its name to one of its residents, François Rakoczy, Prince of Transylvania, exiled in 1711 and received in France by Louis XIV. He opened an «Académie des Jeux» the inspiration for Abbé Prévost's «Manon Lescaut». At the end of the XVIIIth century, Vergennes, Foreign Secretary of State lived there. The decoration of the floors has been conserved. Here, during the Romantic period, Musset and Liszt frequented the literary salon of the Marquise de Blacqueville, the daughter of Maréchal Davout.

Anatole France lived at n°15 from 1844 to 1857. The Hôtel de la Bazinière, or de Bouillon, at n°17 is now occupied by the School of Fine Arts.

It was built by François Mansart in the middle of the XVIIth century for Macé-Bertrand de la Bazinière who received Queen Christina of Sweden there in 1658. It was later bought by Godefroy de la Tour d'Auvergne and Anne-Marie Mancini, Duchesse de Bouillon, a great friend of de La Fontaine who was inspired to write tales based on her love affairs. The state acquired this hotel in 1892. At that time, it belonged to the Caraman-Chimay family. Since then, it has housed the Management and different studios of the School of Fine Arts.

On the main courtyard side, despite the alterations of some details, especially the portal and the wings on the quay, Mansart's original architecture has remained practically unchanged. The façade overlooking the garden with its three dormer windows and its triangular pediment have not been altered.

George Sand lived in a small suite at n°19 between 1832 and 1836; there she wrote «Leila» and frequently received Frédéric Chopin. Anatole France was born in the house in 1844. His father ran a bookshop there.

Buildings on the Quai Malaquais. In the middle, the Ecole des Beaux-Arts.

The quai Voltaire

Begins at the Pont du Carrousel
and ends at the Pont Royal.

Since 1791 it has been named after Voltaire, who died on May 30th 1778 in the Hôtel de Villette situated on the corner of this quay and the Rue de Beaune.

The Hôtel de Tessé, on the corner of the Rue des Saints-Pères, is a beautiful building, whose façade is composed of eight bays of windows separated by monumental pilasters with Ionic capitals. At attic-level, a balustrade runs the length of the building. This mansion was built in 1765-68 by the architect Le Tellier. It has retained its banisters and the splendid decoration of the exhibition rooms. The architect Louis Visconti, Field Marshal Bugeaud, the sculptor Prudon and the archaeologist Barbey de Jouy lived there in the XIXth century.

Buildings on the Quai Voltaire as seen from the opposite bank.

The return of the ashes of Voltaire to Paris in 1792, the procession on the Pont Royal and on the Quai Voltaire where the Church and the Convent of the Théatins can be seen.

At n°9 and n°11 are two mansions built in 1665. The painter Dominique Ingres and the historian Ravaisson-Mollien lived there.

According to an inscription, the newspaper the «Moniteur Universel» was set up at n°13 in 1789.

The neighbouring buildings from n°15 to 25 were built on the site of the ancient Convent of the Théatins. Only a few relics and the south portal of the Convent remain in the Rue de Lille.

N°19, built in the XVIIIth century, was transformed into a hotel for travellers in 1857. There were famous guests such as Baudelaire, Sibélius, Wagner and Oscar Wilde.

The Théatins built the mansions n°23 and n°25, two of the most elegant buildings on the quay, just before the Revolution. Alfred de Musset lived at n°25.

Today all the shops on the Quai Voltaire are occupied by renowned antique dealers or art galleries.

The Hôtel de Ville, whose entrance is situated in the Rue de Beaune, was built in 1661. In 1766 the Marquis de Villette rented it and received Voltaire on his return from Fernay: Voltaire died there

Fireworks and a watershow on the dock of the Louvre in the XVIIIth century. The public on the banks of the Quai Malaquais.

on May 30th 1778. Several rooms have kept their original decoration, especially the reception room.

At n°29 stands the Hôtel de Mailly-Nesle, built in 1632. It has been altered many times since then. It houses the French Documentation services. This mansion is named after Charles Louis de Mailly, Marquis de Nesle. His four daughters, renowned for their beauty, were, successively, the mistresses of Louis XV.

At the corner of the quay and the Rue du Bac stands, on several storeys, a very beautiful building very richly decorated with atlantes and caryatids. It is the site of a house where the famous Charles de Batz, Comte d'Artagnan (1611-1673) lived in the XVIIth century. He was a famous Captain of the Mousquetaires of King Louis XIV and was killed during the siege of Maastricht.

The dock of the Louvre and the Quai Voltaire. A print by Pérelle, XVIIIth century.

The quai du Louvre

Begins at the Pont Neuf
and ends at the Pont du Carrousel

This quay adopted its present appearance under the Second Empire. The part which corresponds to the Palais du Louvre was built by François I, then altered through the centuries.

The shops of the Samaritaine, situated on the corner of the quay and the Rue de la Monnaie were founded in 1870 by Emile Cognacq who chase the name «Samaritaine», as a reference to the famous waterpump that worked on the Pont Neuf between 1609 and 1813.

In 1905, the architect Frantz Jourdain built the first store crowned with domes. It disappeared in 1925 when the architect Sauvage erected the present building facing the quay.

The neighbouring houses have kept their ancient appearance. The building on the corner of the quay and the «Place du Louvre» was erected in 1860 from the plans of the architect Hittorff who redesigned all the monumental buildings located between the quay and the Rue de Rivoli, on both sides of the Saint-Germain l'Auxerrois

The Palais Royal du Louvre at the end of the Middle Ages. Reconstitution by Hoffbauer.

The Saint-Germain l'Auxerrois Church.

Church. It is symmetrical with the one on the corner of the Rue de Rivoli. Its arrangement of Corinthian pilasters resembles the Colonnade of the Louvre.

Lengthy works enabled the gaining of sufficient space in front of the colonnade to create the present «Place du Louvre». The clearing was started by Soufflot, under Louis XVI, continued by Percier and Fontaine under Napoléon I, but was mainly carried out by Haussmann and Hittorff who created the present Place du Louvre.

Out of concern for symmetry and urban splendour, the townhall of the first arrondissement (district), also built by Hittorff between 1857 and 1859, on the left side of the church, was decorated with a façade associating the Gothic and Renaissance styles with a rose window and a pointed pediment above the porch, reminiscent of the architectural layout of Saint-Germain l'Auxerrois.

A Paris-London coaster on the Quai du Louvre, circa 1880. A water-colour by Fraipont.

Built in 1864, the 38-metre high belfry of the townhall has a thirty-eight bell carillon, the largest in Paris. Restored in 1960 this carillon of bells chimes everyday at midday and at 6pm.

The Saint-Germain l'Auxerrois Church was built in several stages between the XIIth and the XVIIth centuries and further modified in the XVIIth century. Devastated by the Revolution and popular riots in 1836, it was entirely restored by the architects de Lassus and Baltard.

It was the parish of the Louvre. Many artists were buried there in the XVIIth and XVIIIth centuries. It contains a large furniture collection, fine quality sculptures and XVIth century stained-glass windows. The choir is surrounded with a beautiful wrought iron railing of the Louis XIV period.

Among the most exceptional elements is a wooden retable carved in Antwerp at the beginning of the XVIth century (1519) representing scenes of the marriage of the Virgin, the tree of Jesse and the Passion of Christ.

The Quai du Louvre in winter, circa 1905. A painting by Galien-Laloue.

The Bassin du Louvre. A painting by Pierre-Antoine Demachy.

A ship on the Quai du Louvre, circa 1920.

THE PALAIS DU LOUVRE

This huge building which stretches eight hundred and sixty metres along the edge of the Seine, with its perimeter of about thirty hectares, holds one of the largest and richest museums in the world.

Here we can only touch on the complex history of this palace and the stages of its construction which cover seven centuries. An extremely important interior restructuration has just been carried out, between 1983 and 1993, creating what is now called the Grand Louvre.

The feudal castle built by King Philippe-Auguste and King Charles V in the XIIIth and XIVth centuries remained until François I and Henry II. The foundations, and in particular those of the dungeon in the west of the Tour Carrée, are still present. The work undertaken in the eighties enabled the restoration and enhancement of these important relics.

It was François I who initiated the reconstruction of the Louvre. The architect Pierre Lescot was asked to draw the plans. Work began in earnest during the reign of Henri II. Pierre Lescot built the southern half of the west wing as far as the Pavillon de l'Horloge.

In 1594, King Henri IV had the main gallery built on the riverside - 560 metres - joining the Cour Carrée to the Palais des Tuileries.

Louis XIV only stayed in the Louvre during the first part of his reign. However, he had a lot of work done on these buildings.

In 1661, construction of the Galerie d'Appolon began perpendicular to the

The Palais du Louvre as seen from the Pont Neuf, circa 1660. An anonymous painting.

The Palais du Louvre as seen from the Quai des Grands Augustins, circa 1640. A print.

Grande Galerie. Its magnificent ceiling was decorated by the painter Lebrun. It was a prefiguration of the Galerie des Glaces of Versailles.

Between 1661 and 1664, the architect Louis Le Vau built the east wing of the Cour Carrée and the south wing, on the Seine side, which was later modified. He began construction of the front of the «colonnade façade».

The main work was completed in 1670. This splendid colonnade, built on monumental foundations, was enhanced by the digging of the ditch in 1964-66 by André Malraux.

The coupled Corinthian columns, the flat pediment, the cornice and its balustrade, the imposing peristyle, all contribute to the majesty and nobility of this masterpiece of classical architecture.

In 1678, once the Court of the King had moved to Versailles, work on the Louvre was again interrupted. The palace continued to serve as a residence for many artists. On the initiative of Charles Perrault, the Academie Française and an exhibition gallery were housed there.

In the XVIIIth century, the palace was gradually transformed into a caravanary. Anarchy ruled to such an extent that artists and writers, among them Voltaire, had to begin a «press campaign» against the scandalous state of disrepair and neglect in which the palace had been left.

Under Louis XV, some restoration was undertaken by Germain Soufflot who also managed to partly clear the surroundings on the Saint-Germain-l'Auxerrois side.

The idea of creating a «museum» at the Louvre was conceived in the middle of the XVIIIth century: a literary man - Lafont de Saint-Yenne - asked that the paintings of the royal collections, especially the ones bought by Mazarin and Colbert, be exhibited in the Grande Galerie.

With the Revolution came new plans for the Louvre. A better link between the Louvre and the Palais des Tuileries was contemplated by the building, on the north side, of a wing more or less symmetrical to the Grande Galerie. In May 1791, the Assemblée Constituante voted by decree the creation of the museum.

The Jean Goujon portal on the riverside of the Grande Galerie.

The loggia of the Louvre on the ground floor of the Henri III wing, under the Galerie d'Apollon.

With the Empire, work on the Louvre was resumed. Napoléon first undertook to evict the artists, including his official painter Louis David. The Cour Carrée was built as much for the architecture as for the sculpture.

Under Louis XVIII, Charles X and Louis-Philippe, the museum was considerably extended.

From 1840 onwards, the newspapers, as in Voltaire's time, asked for the completion of the «Palais historique de la France».

In 1848, the architect Duban undertook various restorations and particularly on the Seine side of the Grande Galerie and the Galerie d'Apollon, whose vault was rebuilt. Delacroix restored the frescoes.

With the Second Empire, considerable means were used. Emperor Napoléon III decided to finish the Louvre. The architect Louis Visconti was chosen to put an end to the works. He began by pulling down the buildings that surrounded the old palace.

Very soon, more than three thousand workers were working on the palace. In December 1853, Louis Visconti died. Hector Lefuel succeeded him. He kept Visconti's general plans, but made many alterations: he added an extra floor to every building.

Lefuel managed to finish this gigantic building site within five years. Extreme care was taken with the buildings which were decorated, painted and carved in profusion. On 14th August 1857 Napoléon III inaugurated the new houses.

On 24th May 1871 the Louvre was spared from the fire of the Tuileries thanks to a few watchful men who managed to contain the blaze during which, however, the library was destroyed. Hector Lefuel devoted the last ten years of his life to repairing the damage and putting the finishing touches to the works.

The most important works were the reconstruction of the Pavillon de Flore, the front of the Grande Galerie up to the booking-offices, and the Pavillon de Marsan towards the Rue de Rivoli. At Lefuel's death in 1880, the palace museum had the appearance that it would retain until 1982.

The plan of the Grand Louvre.

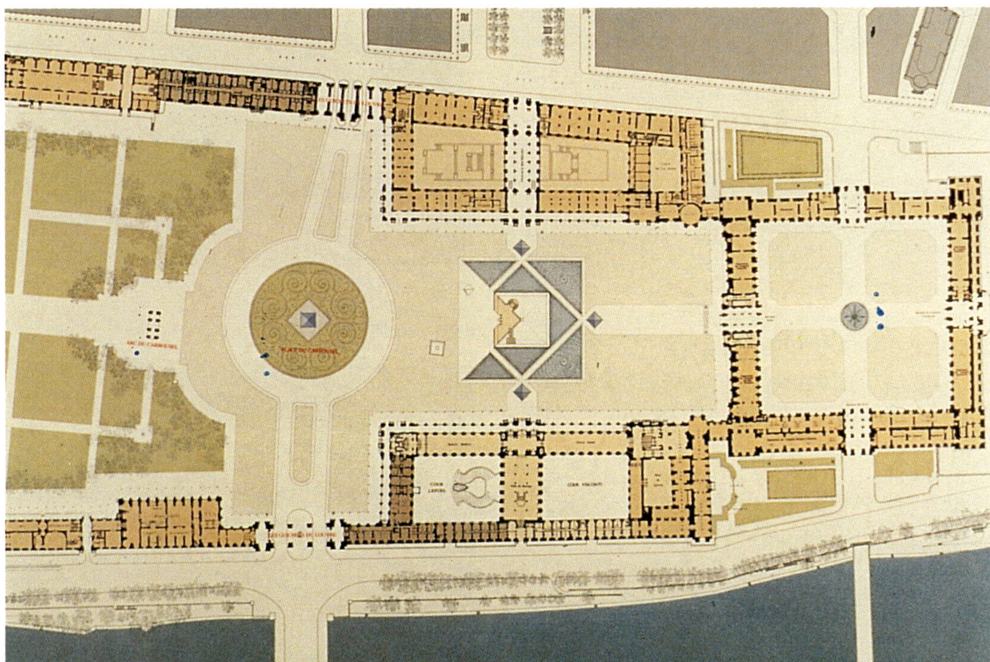

A perspective view of the front of the Grande Galerie.

The booking offices of the Louvre.

The important works under the Cour Napoléon between 1983 and 1993, carried out under the supervision of the architects Ieoh Ming Peï and Michel Macary, enabled a thorough reorganization of the galleries.

Since its inauguration in Spring 1985, the transparent Peï pyramid has been the principal entrance to the museum. It leads to a magnificent reception hall. Together, they are considered among the best architectural achievements of our time.

The inauguration of the pyramid, its ornamental pools and its fountains in the main Napoléon III courtyard provided the opportunity to place a lead copy of the equestrian statue of Louis XIV which had been designed for Versailles by Le Bernin.

The departure of the «Ministère des Finances», housed in the Louvre for about one hundred and twenty years, enabled the clearing of large areas and made available to the public the wonderful Napoléon III decoration in the Richelieu wing and particularly the sumptuous showrooms of the Pavillon Turgot, the banquet room and the grandiose neo-Classical and neo-

The Cour Napoléon III and the Peï Pyramid.

Renaissance staircase, which were restored in 1930.

1993, which marked the tricentenary of the official creation of the Museum National in the royal palace of the Louvre, coincided with the completion of the works of the Grand Louvre. The collections are now presented in new thematic, historical and more logical arrangements. The surface area for exhibitions has doubled. Now, more than ever, the Louvre merits its title of the greatest museum in the world.

(*) Musée du Louvre : open every day except on Tuesdays from 9am to 6pm; late opening on Wednesdays till 9.45pm. Tel: 40 20 50 50. Entrance through the Pyramid.

(*) Musée des Arts Décoratifs open every day except on Tuesdays from 12.30 pm to 6 pm. Tel: 42 60 32 14. Entrance through n° 107 Rue de Rivoli.

A Pavilion of the Cour Carrée, Colonnade wing (XVIIth century).

The pont Royal

Built entirely in freestone, the Pont Royal, with its humback arch, is, with the Pont Neuf and the Pont Marie, one of the three oldest bridges in Paris. It has five semicircular arches of unequal span. It is 17 metres wide and 110 metres long. Splays make access from the quays easier.

It is the only bridge in Paris not in the axis of the roads to which it gives access. On the right bank, it is opposite the Pavillon de Flore, on the left bank it opens out after a bend on the Rue du Bac.

It was built between 1685 and 1689 from the drawings of three architects: Jules Hardouin-Mansart, Jacques Gabriel and the Dominican François Romain.

With the Pont Royal, the links between the Faubourg Saint-Germain and the Louvre became very convenient, the coaches and the horsemen no longer had to make a detour over the Pont Neuf.

The Pont Royal was, for a long time, the only bridge downstream of the Pont Neuf. It was only in 1788 that the Pont de la Concorde was built to ease the traffic between the Faubourg Saint-Germain and the Faubourg Saint-Honoré, and under the first Empire that the Pont d'Iéna allowed access between the Champ de Mars and the colline de Chaillot.

The Pont Royal in the XIXth century. A painting by Harpignies.

The Pont Royal and the Bassin du Louvre in the XVIIth century. A painting by Gravenboek, Carnavalet.

The Pont Royal, upstream and downstream.

The ferry and the pont Royal

From the earliest times, the first and most common way of crossing streams and rivers, that could not be forded, that is to say on foot or on horseback, was the ferry. For centuries these ferries were numerous in Paris and in the suburbs where bridges were rare until the XVIIth century. Outside Paris, there existed only a few bridges before the end of the XVIIIth century, such as that of Charenton on the Marne, the very old Sevres and Saint-Cloud bridges, and that of Neuilly. Elsewhere crossing was by ferry. These ferries carried pedestrians, horses and teams from one bank to the other. The Bac des Tuileries, which disappeared when the Pont Royal was completed in 1689, was used to transport water to the men who carted the stones, necessary for the buiding of the palaces and edifices on the right bank, from the quarries of the village of Vaugirard.

The buildings of the Rue du Bac were built in the XVIIth and XVIIIth centuries along this ancient route.

Ferries were generally guided by a ferry cable, a cable stretched from one bank to the other, making the crossing easier and preventing them from being carried off course by the current.

Ferries were sometimes replaced by wooden bridges or ran alongside them. This was the case for the ferry of the Tuileries. A timber foot-bridge, preceded the masonry bridge built by Louis XIV. This foot-bridge was called «the Pont Rouge» in contemporary documents because of the red lead paint with which it was painted. Built in 1632 it collapsed in 1864, which led to the building of the bridge.

The quai des Tuileries

Begins at the Pont du Carrousel
and ends at the Pont de la Concorde.

This quay took its definitive appearance after the Second Empire. It was planted with a row of plane-trees, as were the other quays in Paris. Before Haussman, there were no trees except in the Cours la Reine.

Napoléon I decided, in 1806, to replace the roadway that was below the terrace of the Jardin des Tuileries by a stone quay.

The long terrace with its balustrade overlooking the Quai des Tuileries, is 800 metres long. It is called «the Waterside Terrace» because of its position. The «Terrasse des Feuillants» which is symmetrical to it along the Rue de Rivoli is named after the Convent of the Feuillants that stood nearby before the French Revolution (rue Castiglione).

Between the Flore and Marsan pavilions which end the Palais du Louvre on the west side, the magnificent Palais des Tuileries stood along 330 metres. It was burnt down by the Commune in May 1871.

The Quai, the Jardin and the Palais des Tuileries in 1806.
A painting by Bouhot, Carnavalet.

The Pavillon de Flore in 1890. A painting by Firmin-Girard.

The ground on which, from 1564 onwards, Queen Catherine de Médicis had the first Palais des Tuileries built, was outside the walls of Paris. The surrounding Charles V wall, rediscovered during the works on the Grand Louvre in 1991-92 is slightly in front of the Arc de Triomphe du Carrousel.

The architect Philibert de l'Orme built the first Palais des Tuileries. It has since been considerably enlarged and altered on several occasions, notably by Louis Le Vau, François d'Orbay and Lemercier in the XVIIth century and decorated at that time by the greatest artists who also worked on the Louvre and on the Château de Versailles.

In the XIXth century, this palace lived its most magnificent moments during the reigns of Napoléon I and Napoléon III.

Ten years after its burning down by the Commune, the ruins were still there and architects pro-

Le Nôtre French Garden

The «Jardin des Tuileries», like those of the Louvre and the Carrousel, is a museum of open air sculptures that has been constantly enriched, since the XVIIth century, with many works being placed in the XIXth century and in modern times. There are three important works on the terrace on the riverside. On either side of the Avenue du Général Lemonier near the «Pavillion de Flore», there are two stone female sphinxes, which were taken from the town of Sébastopol, after the victory of the French in 1856. On the terrace, not far from the access stairs near the entrance of an imposing underground telephone exchange set up in 1965, stands a large bronze group by Paul Landowski finished in 1906 and representing the Sons of Caïn in a dramatically realistic posture. Near the buildings of the Orangerie there is a lion struggling with a snake, a moulding from an original bronze by the sculptor Barye. At the end of the balustrade overlooking the quay and the Place de la Concorde, is a superb stone lion by the Florentine sculptor Pierre Tacca (XVIIth century). A lion of the same style by the same artist is placed on the terrace on the side of the Rue de Rivoli.

A stone lion by the sculptor Tacca, early XVIIth century, on the corner of the Quai des Tuileries.

posed many restoration projects, but the government of the Third Republic decided, in 1882, to have them pulled down...

At the same time as the construction of the Palais des Tuileries, Catherine de Médicis had a garden created. In the middle of the following century, André Le Nôtre transformed it entirely. He removed the quincuces and the primitive borders and also the numerous residences that were there. Grandson and son of royal gardeners, André Le Nôtre was born in the Tuileries and had always lived there in a house adjacent to the Palais. A bust, near the large circular pool, honours his memory. He built the long terrace overlooking the Seine . This was one of the most beautiful walks in Paris overlooking the Seine. He set out the horseshoe-shaped access ramps which encircle the large pool and end on a wide central space with a view to the horizon.

Having redesigned the garden and its views of elm-trees and pools from the central Pavillon du Palais, Le Nôtre designed and planted the grand avenue up to Chaillot Hill: this became the Avenue des Champs Elysées.

Perspectives in the «Jardins des Tuileries» near the large circular pool.

The Orangerie (Orangery) of the Jardin des Tuileries matches the Jeu de Paume building. They both contain museums and temporary exhibition rooms. Extensive works were carried out on the Musée de l'Orangerie(*) in the 1980s. The rooms on the ground floor exhibit large mural paintings by Claude Monet, painted between 1890 and 1921 at Giverny, and known as the «Nympheas» (water-lillies).

(*) Museum open everyday from 9.45am to 5pm except on Tuesdays. Tel: 42 97 48 16.

The archaeological excavations undertaken in 1993 led to the discovery of the foundations of different buildings dating from before the Le Nôtre French Garden, notably the «Polygon» built for the training of young King Louis XIII.

The quai Anatole France

Begins at the Pont Royal
and ends at the Pont de la Concorde.

The name of the writer Anatole France was given, in 1947, to this part of the ancient Quai d'Orsay which was built as a tree-lined walk in 1708 by the Prévost des Marchands, Charles Boucher d'Orsay.

The site, occupied by the buildings of the «Caisse des Dépôts et Consignations» and the «Musée d'Orsay», was, in the XVIIth century a wood store.

On the Rue du Bac side, there were two beautiful mansions built by the architect Robert de Cotte in 1727. These mansions and the neighbouring houses were burnt down by the Commune. In the courtyard of the Caisse des Dépôts, there is still a pediment from the ancient Hôtel de Bel-Isle, which depicts Minerva looking at architectural plans.

The present building was erected in the 1880s. Two motifs are sculpted on the canted wall on the side of the Rue du Bac. The one on the upper pediment is an allegory of Economy.

On the site of the Musée d'Orsay was the Palais d'Orsay built between 1810 and 1838 for the «Affaires Etrangères» (French Foreign office)

Detail of the sculptures crowning the Musée d'Orsay.

The Quai d'Orsay at the end of the XVIIIth century. Anonymous.

On the parvis of the Musée d'Orsay, a horse rearing above a harrow, 1878. Sculpture by Rouillard.

The Musée d'Orsay

The architecture of this building is a very interesting model of the 1900s. The stone façade overlooking the quay hides a huge nave covered with elegant iron railings. This nave is 140 metres long, 40 metres wide and 50 metres high. On the main façade, level with the roofs, are allegorical statues by Dominique Hugues, Laurent Marqueste and Antoine Enjalbert. They represent the towns of Bordeaux, Toulouse and Nantes under the forms of young seated women, holding coats of arms.

On the parvis towards the Rue de Bellechasse statues have been placed that were in the Jardins du Trocadéro from 1878 to 1936: a rearing horse above a harrow by Pierre Rouillard, a rhinoceros by Henri Jacquemard, an elephant by Emmanuel Frémiet, a statue of Europe by Alexandre Falguière, Africa by Eugène Delaplanche, North America by Eugène Hiollé, South America by Aimé Millet and Oceania by Mathurin Moreau.

occupied by the «Cour des Comptes» and the «Conseil d'Etat». It was burnt down by the Commune in May 1871.

Here, in 1897, the «Compagnie des Chemins de Fer d'Orléans» (Orléans Railway Company) commissioned the architect Victor Laloux to design the monumental Orsay Station and a six hundred room palace, named the Hôtel d'Orsay. These refurbished buildings now house the collections of the Musée d'Orsay.

It contains very important collections of objects of art, paintings and sculptures representing various trends and «French Schools» of the XIXth century, an age to which it is exclusively devoted. It also organizes, several times a year, exhibitions with themes on architecture, decorative arts and large retrospectives on the artists and the European schools in the XIXth century.

From the top-floor terrace overlooking the Seine, visitors on the right bank have a view from the Louvre to Montmartre and beyond.

The grandiose banqueting hall of the ancient Hôtel d'Orsay and its painted and carved allegorical decoration have been retained. They serve as the setting for the Museum restaurant.[1]

The Hôtel de Salm, next to the Orsay museum, was built in 1782 and 1790 by the architect Pierre Rousseau for the Prince of Salm-Kyrburg.

The raising of the Quay in the XIXth century threatened the visual harmony on the side of the Seine. Originally there was a terrace garden stretching down to the banks. The Hôtel de Salm was

Orsay station and passenger riverboats in 1900.

The Orsay Museum as seen from the Quai des Tuileries.

bought in 1804 by the «Grande Chancellerie de la Légion d'Honneur». It was entirely restored after the great fire caused by the Commune.

The exterior sculptures were spared from the fire. There are many fine sculptures on the quay. 19 busts of ancient characters by the sculptor Philippe Roland, are placed in niches. The six allegorical statues which dominate the cornice are by Jean Moitte, as are the five low reliefs placed above the entablatures of the windows of the central semicircle. Above the windows, the side façades are decorated with 50 low reliefs representing allegorical scenes. They are the work of the sculptor Clodion.

Towards the Rue de Lille, the main portal is topped by two statues of Fame by Jean Moitte. The sallies are decorated with low reliefs in an antique style by Philippe Roland.

From this portal, a triumphal arch, lead lateral colonnades matching a monumental peristyle composed of six Corinthian columns which leads to the Salons of the Légion d'Honneur. The interior was redecorated in 1874. There are two

The Palais Bourbon.

double-scale replicas of this palace, one built at Saint Arnoult en Yvelines by the diamond merchant Porgès at the beginning of the XXth century, the other in San Francisco by a rich collector passionately interested in French architecture...

The museum of the «Légion d'Honneur et des Ordres de Chevalerie» occupies most of the galleries of the Hôtel de Salm. The entrance is at n°2 Rue de Bellechasse. This museum, created in 1925 by General Dubail, displays a large collection of decorations, badges and medals illustrating the ancient Orders of Chivalry and also pictures, objects of art and arms. A gallery is devoted to the history of the Order of the Legion of Honour, created by Napoleon in 1802 to honour the military and civil services paid to France.

The building, at the corner of the Rue Solferino is a fine neo-Classical construction from the Second Empire, whose decorations and rooms have been retained.

The Deligny public baths

There had been public baths in Paris since the Middle Ages. In the XIIIth and XIVth centuries, they were called «steam rooms». In 1761, a barber and owner of a bath-house, named Poitevin, set up the first public hot bath-house on a boat moored near the Tuileries. Its immediate success led to the opening of a second near the quai d'Anjou. Cold baths appeared too and «Chinese» baths were in fashion. Two were reported in 1785 near the quai de la Tournelle. In 1801, Turquin, the inventor of these swimming schools, joined up with his son-in-law Deligny, and built large construction based on a basin surrounded with beaches and

The palace of the Legion of Honour and its copper dome.

The Deligny swimming pool and the buildings on the Quai Anatole-France.

Beyond, are the façades overlooking the garden of the Hôtel Beauharnais, today the residence of the Ambassador of the Federal Republic of Germany, built by Boffrand and redecorated under the Empire, as well as the Hôtel de Seignelay, also built by Boffrand in 1714, and currently occupied by a Minister.

N° 25 is a beautiful neo-Palladian mansion decorated with columns and sallies. The façade, which is set back, is preceded by a perron with two large statues draped in an Antique style.

Buildings n°27 and n°29 are indicative of the architectural research of the 1900s and of the Art Nouveau era. They were built in 1905 by the architect Bouwens van der Boijen.

The quay ends with a beautiful rounded building built in the Second Empire.

1. Open every day at lunchtime except on Mondays and also open on Thursday nights. Entrance on the parvis at n°7 Rue de Bellechasse. Tel: 45 49 42 33.

The main entrance of the Orsay museum is on the side parvis. Entrance for special exhibitions is via a specific entrance on the Quai d'Orsay. Open every day except on Mondays from 10am to 5.30pm, on Thursdays to 10pm. Tel: 40 49 48 14 / 45 49 11 11.

peripheral walkways where the cabins were placed.

In 1808 Deligny created his own swimming school near the Pont de la Concorde, towards the Quai d'Orsay, where it was rebuilt and modernized under the name «Bains Deligny».

Up to July 8th 1993, the date of the demolition of this establishment, it was the last construction of this kind on the Seine. It therefore presented a real historical interest. It had up to 340 cabins and also offered public or private rooms, a café, a restaurant, hairdressers' and chiropodists' salons, all on two storeys .

In 1847 the barge which had transported the ashes of Napoleon, was used for the basin. In 1919, the Richard family bought it. In 1977 a finance company was entrusted with its destiny under the management of Philippe Richard. While this guide is under press, the reconstruction of the Deligny baths is being considered.

The ancient pont de Solférino

Before 1858, the waterside terrace, a superb viewpoint over the Seine, nearly 1 kilometre long, conceived by Le Nôtre while he was redesigning the Jardin des Tuileries, was still such as the landscape gardener had designed it. It was a barrier to a transversal passage linking the Rue de Castiglione to the Quay, a barrier nobody had dared touch until then and which was also an obstacle to the presence of a new bridge connecting the Tuileries to the left bank.

In 1858, Emperor Napoleon III allowed this opening, compensated in those times by the presence of a wide foot-bridge linking the two sections of the terrace. The «Administration des Ponts et Chaussées» (the Department of Civil Engineering) seized this opportunity to build a new bridge, all the more important as the start of the Boulevard Saint-Germain was under construction and a throughlink seemed necessary.

The bridge was inaugurated after Solferino's victory over the Austrians (in Lombardia).

This bridge had three depressed elliptical arches with a clear span of 40 metres resting on masonry piers. They supported an iron deck of 20 metres.

Romany and de Lagallisserie were the engineers and designers of this bridge. This fine work was pulled down in 1960. A «temporary» foot-bridge replaced it until 1993.

The ancient Pont de Solférino and the bank of the Quai d'Orsay (Anatole France). A water-colour by Fraipont.

The passerelle Solférino

The project presented by Marc Mimram in 1992 is to be built in 1993-1994. It is a work of a rare elegance and it is particularly original in its architectural and technical design. Miriam is both the engineer and the architect.

Like the Pont Alexandre III, the Passerelle Solférino crosses the Seine with a single arch, which spans 110 metres, centred exactly on the axis of the river and thus leaves a clear view of the Pont de la Concorde downstream and of the Pont Royal upstream. The orignality consists of giving two access levels to the foot-bridge: upper access on the right and left banks level with the present pavements, and access halfway between the pavement and the banks in line with the pedestrian tunnel of the quay on the right bank, which leads to the Jardin des Tuileries.

Pedestrians arriving at this mid-level will therefore have in front of them the sight of the structure of the arches and of those of the ribs and crossbeams bearing the deck of the footbridge, which means that, at last pedestrians will have access to the somewhat fascinating sight of the under side of an iron bridge. They will progress inside the structure itself along two slightly sloping symmetrical ramps.

There will be therefore, in each direction, two possible passages from the left and right banks and from the middle of the foot-bridge where both passages meet along several dozen metres.

The transparences, the cast shadows and the light patterns will thus be combined in some sort of play starring only the bridge and the river. The foot-bridge will be lit at night by a system integrated into the structure, which will outline it after dark and create a magical visual show suspended in the skies over Paris.

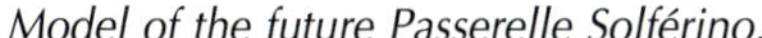

Model of the future Passerelle Solférino.

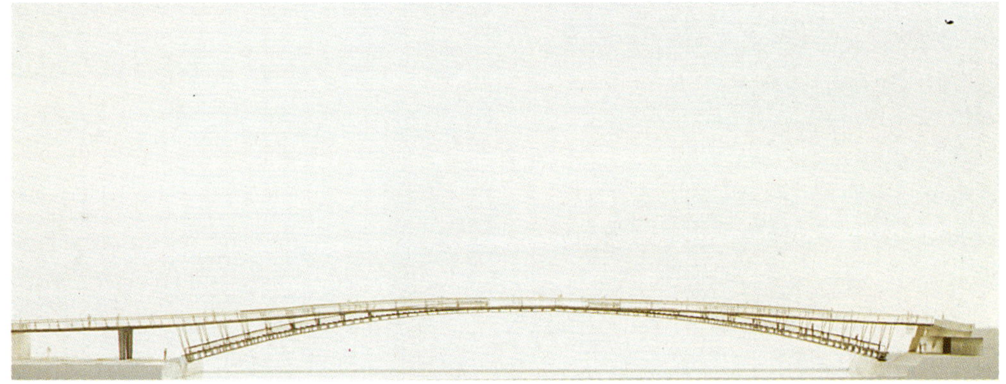

The pont de la Concorde

The Pont de la Concorde is the work of the engineer, Jean Rodolphe Perronet (1708-1794), founder of the Ecole Nationale des Ponts et Chaussées. He was also the designer of the old bridges of Neuilly, Melun, Mantes, Pont Sainte-Maxence, etc.

The Pont de la Concorde was built in 1787-1791, with the help of the engineer Demoustier, with materials which came partly from the pulling down of the Bastille. It was called Pont Louis XVI, then de la Concorde under the Consulate, then again Louis XVI at the time of the Restoration and finally «de la Concorde» in 1830.

In 1810, Napoleon ordered the statues of twelve generals killed during the Empire Campaigns to be erected on pedestals placed above the piers.

In 1828, they were replaced by the statues of twelve prominent characters: four great ministers: Colbert, Richelieu, Suger and Sully; four generals: Bayard, Condé, Duguescli and, Turenne; and four sailors: Duguay-Trouin, Duquesne, Suffren and Tourville.

The weight of these colossal white marble statues endangered the bridge. They were transferred to the «Cour des Invalides», then to Versailles where they remained for a century. Those which have been kept are today in the cities where these great men were born. Those of Duguay-Trouin, Tourville and Duquesne stand in the main courtyard of the Naval College of Coetquidan.

The width of the Pont de la Concorde was doubled in 1930-32 with the addition of spans downstream and upstream.

This widening was undertaken with great sensitivity, respecting the profile of Perronet's bridge. Today the work measures 34 by 153 metres with a marked humpback arch.

The Pont de la Concorde. A painting by Zuber.

Sous le pont de la Concorde.

Restaurants on the riverside

The restaurants on the riverside are few, undoubtedly due to the competition with the restaurant-boats of the main tourist navigation companies which all offer specialized cruises with dinner or lunch on board. However from upstream to downstream, there are still some spaces where charm and the unexpected mingle well. They are:

● The restaurant ***«Le grand Bleu»*** on the Port de l'Arsenal which enjoys a pleasant location overlooking the garden and the basin. Tel: 43 45 19 99.

● The restaurant-boat ***«Abou Dahou»***, on the bank of the Quai de New York. Tel: 47 23 62 96.

● The restaurant-barge ***«Le Pacific Louisiane»*** is also used as a pontoon for the «Compagnie des Vedettes Paris Ile-de-France» on the Suffren Port. Tel: 47 05 71 29.

● Nearby in the same port, the restaurant-boat ***«Le Mariachi»*** is a large boat with two rooms. It organizes dinners accompanied by a Mexican orchestra. Tel: 45 55 43 67.

● Between the Ponts de Grenelle and Mirabeau on the Quai de Grenelle, Port de Javel, the restaurant ***«La Plage»*** - Base Alpha enjoys a fine southwest location. Its tables are placed on the wide bank in the shade of elegant canvas awnings.

● ***«Le Princess Elizabeth»***, a paddle steamer burdened with history, moored on the quay under the Pont Mirabeau, now very Parisian, welcomes visitors on board for festivals, events, exhibitions and cocktails. Built in 1926, at the time of the birth of the future Queen of England, it was originally intended to transport passengers between Southampton and the neighbouring islands and later became a minesweeper, a troop carrier, decorated for feat of arms, the hero of a film by Walt Disney *(the Children of Captain Grant)* and a cruise boat. Restored in its original style by the association for the Defence of Typographic Art, it combines English refinement and French gastronomy with a wonderful view of the statue of Liberty and the Eiffel Tower. Base Alpha, Port de Javel Haut, 75015. Tel: 40 59 01 71.

● Close to Paris, near the Pont d'Issy, the ***«River Café»*** barge has been berthed for several years. This restaurant where the jet set meets the World of Press and Advertising, is always crowded. River Café, 146 quai de Stalingrad, Issy-les-Moulineaux. Tel: 40 93 50 20.

The place de la Concorde

The Obelisk, Place de la Concorde.

The layant out of this square was achieved under Louis XV whose name it bore until the Revolution.

Its present name was given in 1830 at the beginning of the reign of Louis Philippe.

In 1763 the sculptor Edme Bouchardon erected, in the centre, an equestrian statue of Louis XV which was destroyed during the Revolution.

The square was designed by Jacques-Ange Gabriel who built, between 1760 and 1775, the two symmetrical palaces on either side of the Rue Royale.

Each of them is composed of a large portico delimited by twelve Corinthian columns and bordered with two sallies.

The Place de la Concorde as seen from the ancient Ministère de la Guerre (War Office).

The building on the right was allocated to the «Garde-Meubles National», then, under the Convention to the «Ministère de la Marine» which is still there.

The building on the left contains several private mansions, that of the Automobile Club de France is in the centre. The Hôtel du Duc D'Aumont was bought by the Comte de Crillon, a descendant of Crillon le Brave, a brother-in-arms of Henri IV. At the end of the XIXth century, it belonged to the Duchesse de Polignac, the daughter of the Marquis de Crillon. It was acquired in 1907 by the company of the «Grands Magasins et Hôtels du Louvre» which set up the palace, the Hôtel de Crillon, inaugurated on 11th March 1909.

The square is decorated with numerous sculpted monuments. The great fountains made from heavily sculpted superposed basins are dedicated one to the river deities and the other to the sea deities.

The architect Hittorff also designed, for this immense square which covers almost eight hectares (360m by 210m), splendid rostral columns with, on their upper pars, half naves supporting the lantern. These twelve magnificent columns are currently being restored.

Eight small symmetrical pavilions, linked by balustrades, which formerly overlooked ditches which were filled during the reign of Louis Philippe, support the statues symbolising eight cities: Brest and Rouen by Cortot, Lyons and Marseilles by Petitot, Bordeaux and Nantes by Caillonette and Lille and Strasburg by Pradier.

Some of the marble horses sculpted for the watering place of Marly-le-Roi by Coysevox and Coustou were placed, in the XVIIIth century, at the entrance of the Tuileries, the others at the entrance of the Champs Elysées.

The sculptor Michel Bourbon has made replicas of them, the originals now being in the sculpture galleries of the Nouveau Louvre.

Place de la Concorde, the bridge and Palais Bourbon.

The Obelisk

The Luxor obelisk was offered to France by the viceroy of Egypt Mohammed Ali. It was erected on the square on 25th October 1836. It came from the temple of the Pharaoh Ramses II in Luxor, a long and difficult journey which took almost three years. The obelisk, placed on a high pedestal, is 22.83 metres high and weighs more than 220 tonnes. Three hundred thousand people attended its erection.

The cours la Reine

Begins at the Pont de la Concorde
and ends at the Pont de l'Alma.

In 1616, Queen Catherine de Médicis commissioned the achievement of this avenue, a long tree-lined walk of 1500 metres, closed at its ends by monumental railings. The Cours was a very popular walk until about 1775, then little by little deserted in favour of the «Grand Cours»: the Champs-Elysées.

Two large palaces, the Grand and Petit Palais, were built on the edge of the Cours la Reine on the occasion of the 1900 World Exhibition.

Their façades -unique in Paris- do not face the Seine. They stand on either side of the avenue that links the Esplanade des Invalides to the Champs-Elysées, across the Pont Alexandre III.

The Cours la Reine as seen from Chaillot in the middle of the XVIIth century.

The 1855 World Exhibition, the Galerie des Machines on the Cours la Reine.

The Grand Palais, erected on a four-hectare plot was built by the architects Deglane, Thomas and Girault. This huge building combines two architectural styles and two building techniques. The neo-Classical ashlar facades, heavily decorated with mosaïc and ceramic friezes and allegorical sculptures, conceal a huge steel and glass architecture, 180 metres long and 48 metres high. At the edges of the main façade, on the frontons which crown the side entrances, were placed the two immense allegorical equestrian groups by the sculptor Georges Recipon: «Immortality preceding Time» on the side of the Champs-Elysées and «Harmony downing Discord» on the side of the Cours la Reine.

Several dozen allegorical sculptures, a huge mosaïc: «Art through the Ages» by Joseph Blanc, and an exceptional high-relief ceramic representing «the Stages of Civilization» by Joscph Blanc, decorate the façades of this monument, one of the largest in Paris.

The back wing of the Grand Palais, facing the Avenue Franklin Roosevelt has housed the «Palais

At the far end of the vista towards the Place de la Concorde, the equestrian effigy of King Albert I of Belgium is a work by Armand Martial erected in 1938.

Equestrian statue of Albert I.

The flood-lit Grand Palais

The Grand Palais

The great hall has been devoted, since 1940, to the organisation of all kinds of shows and exhibitions. Nowadays, it is events such as the Salon du Livre, the Biennale des Antiquaires and the FIAC which attract the most people. The Motor Show was created at the Grand Palais in 1906.

The galleries on the Champs Elysées side are used for temporary exhibitions which are generally highly successful, such as the retrospective «Toulouse Lautrec» in 1992 and «Titian and Amenophis III» in 1993. Great thematic exhibitions also take place at the Petit Palais.

de la Découverte» and the Planétarium since the 1937 World Exhibition.

The Petit Palais, built by the architect Charles Girault, is also a large neo-Classical building with a trapezoidal layout, the four main parts of which surround, in the centre, a large patio decorated with a peripheral colonnade of pink granite.

The central portal of the Petit Palais and the two symmetrical wings of the main façade as well as the whole of the building were provided with an abundant sculpted decoration; the exhibition rooms and the main hall are decorated with immense allegorical frescoes. This palace, essentially meant to house the painting and sculpture collections of the Ville de Paris, also contains collections of objects of art and furniture.

On the Cours Albert I, named in 1918 in honour of the King of Belgium, numbers 18, 32, 34, 36 and 40, formerly the town houses Jules

Ferry, Boselli, La Ferronnaye, Schneider, Fould and de Vibraye, are fine constructions from the second half of the XIXth century, interesting testimonies to the so-called eclectic style that characterized that time. Some of them still have their magnificent interior decoration.

N°26 has a neo-Classical appearance with Ionic pilasters and sculpted motifs.

The Fould and Schneider town houses, numbers 32 and 34, house the Brazilian Embassy.

The private mansion of the master glass-maker and decorator René Lalique at n°40 presents an important façade surmounted by pinnacles and a portal with floral motifs.

On the Cours itself a certain number of high quality monumental sculptures have been placed.

Near the Place de l'Alma stands the monument to the Polish poet and patriot Adam Mickiewicz, completed by Antoine Bourdelle in 1929.

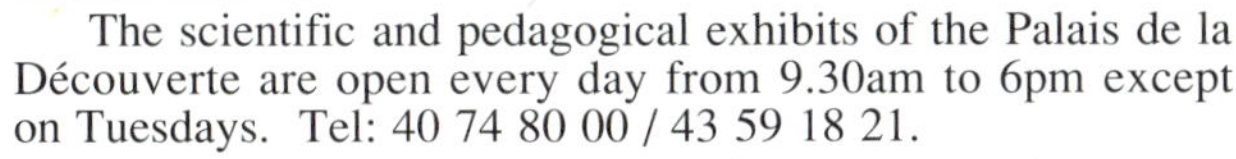

The scientific and pedagogical exhibits of the Palais de la Découverte are open every day from 9.30am to 6pm except on Tuesdays. Tel: 40 74 80 00 / 43 59 18 21.

The museum of the Petit Palais is open every day except on Tuesdays from 9.30am to 6pm.

The main façade of the Petit Palais.

Equestrian statues of Simon Bolivar by Emmanuel Fremiet.

Bolivar and La Fayette

Near the Place de l'Alma stands the monument to the Polish poet and patriot, Adam Mickiewicz, completed by Antoine Bourdelle in 1929.

On the lawns of the Cours la Reine three magnificent equestrian statues have been placed.

On each side of the Alexandre III bridge, and facing each other, are the two statues of men who fought for the freedom of nations at the end of the XVIIIth century: Generals de La Fayette and Simon Bolivar. The equestrian statue of La Fayette was achieved by Paul Barthet, a pupil of Frémiet, and offered to France in 1906 by the school children of the United States. The statue of Simon Bolivar was achieved by Emmanuel Frémiet in 1933: it was offered to the Ville de Paris by the Bolivarian republics of Venezuela, Columbia, Ecuador, Peru, Bolivia and Panama.

The quai d'Orsay

Begins at the Pont de la Concorde and ends at the Pont de l'Alma.

This quay acquired its present form under the First Empire. Intensive working of the Seine river bed and the linking up of the ancient Ile des Cygnes to the left bank, completely altered the banks between the Invalides and the Champ de Mars.

Opposite the Pont de la Concorde, the Assemblée Nationale or the Palais Bourbon is the first of the large buildings on the quay. It was built on the site of the Hôtel de Bourbon, itself erected for Louise Françoise de Bourbon, the daughter of Louis XIV and Madame de Montespan. During the Revolution, the Hôtel de Bourbon was seized from the Princes de Condé.

In 1795, the Conseil des Cinq Cents (the Council of the Five Hundred), the first Assemblée Nationale, was established there. The architects de Gisors and Lecomte built a Council Chamber. In 1806, the architect Poyet built the façade, whose Corinthian colonnade supports an imposing sculpted pediment.

Steamer facing the Quai d'Orsay in the middle of the XIXth *century.*

The Palais Bourbon and the Quai d'Orsay in 1840.
A water-colour by Fraipont.

Under the Restoration, the architect Jules de Joly rebuilt the Council Chamber. A large library was set up there. This spacious domed gallery was heavily decorated with frescoes by, in particular, Eugène Delacroix.

The Hôtel de Lassay, set back from the quay and facing a large terraced garden was built for the Marquis de Lassay from the plans of the architect Lassurance in 1722-1724.

From 1794-1804 it housed the Ecole Polytechnique which had just been created by Napoleon. In 1843 it became the residence of the Presidents of the Chamber of Deputies.

Republican statuary

The large triangular pediment of the Palais Bourbon is composed of a high relief, achieved in 1842 by the sculptor Jean-Pierre Cartot. It represents «France», placed between «Liberty» and «Law and Order» with the Spirits of Trade, Agriculture, Peace, War and Eloquence. On the wings which are set back from the pediment, the two large high reliefs represent «Prometheus livening the Arts» by François Rude, on the right and «Public Education» by James Pradier on the left. The tall statues standing on the perron are, on the right, «Themis» by Antoine Houdon and on the left, «Minerva» by Philippe Roland. On the pedestals, in line with the railings, have been placed the statues of four great politicians and humanists devoted to the grandeur and dignity of France: Maurice de Sully, Michel de l'Hospital, Henri-François d'Aguesseau and Jean-Baptiste Colbert.

Monument to Aristide Briand.

Monument to Aristide Briand

On the quay, built into the long railings which protect the main courtyard, was erected, in 1937, the first monument by the sculptor Paul Landowski to the memory of Aristide Briand (1862 - 1932). This great orator, many times «Président du Conseil», Ministre des Affaires Etrangères (Foreign Secretary) fought for Franco-German Peace. This beautiful bronze group evokes Labour in the Fields made possible by Peace.

The Grand Gallery of the Hôtel de Lassay.

The rooms and the reception rooms have kept their rich and refined XVIIIth century ornamentation. The large «Salle des Fêtes» was built under the Second Empire by the Duc de Morny. It joins the Hôtel de Lassy to the Palais Bourbon.

The magnificent «Palais des Affaires Etrangères» is near the Hôtel de Lassay. Its construction began in 1845 and it was completed in 1867. The architect Jacques Lacornée was inspired by the Classical styles of the XVIIIth Century. The façade has two series of columns, Doric on the ground floor and Ionic on the upper floors. The central part is framed by two pronounced sallies which reinforce the monumental character of the building. Richly decorated oval medallions are placed above the upper-floor windows.

The main rooms as well as the large banqueting hall and the state rooms are decorated in the neo-Classical eclectic style, inspired by Versailles.

The construction of the Invalides («the greatest thought of my reign» said Louis XIV) began in 1671. The minister Louis Louvois was entrusted with the supervision of the works. He chose the architect Libéral Bruant. His talent was exceptional, and his attention to detail was such that, the hospital-hospice, begun in 1671, was ready to accomodate the first disabled people in 1674.

Designed to house 4,000 «Caducs» soldiers, that is to say wounded in battle, the building represents the strength and magnificence of the King's intentions. The façade, 224 metres wide, is the most homogeneous and grandiose in Paris, the main courtyard, harsh but majestic, the two churches; the Soldiers' Church and the Dome Church, the sixteen inner courtyards around which the immense constructions rise up in an irreproachable order; all reflect the desire for grandeur. For three hundred years, this building has never stopped being used as a hospital institution although this function has been progressively reduced in favour of the immense collections of the Musée de l'Armée (Army Museum) and the Musée des Plans-Reliefs (Model Gallery)[1].

The Pediment of the Invalides

In 1733, Louis XV entrusted Guillaume Coustou with the achievement of the grand equestrian high relief representing Louis XIV, on the tympanum of the central doorway; destroyed during the Revolution, this high relief was repaired by Cartellier in 1815. The Dome Church, the largest religious building of the reign of Louis XIV, was achieved by Jules Hardouin Mansart between 1680 and 1706 and completed by Robert de Cotte.

A circular room of the Ministère des Affaires Etrangères (Foreign Office).

The Esplanade of the Invalides during the 1900 World Exhibition.

Beyond the Esplanade, the Quai d'Orsay has a series of high buildings from various ages. In February 1993, on the flowerbed which is at right angles with the Pont des Invalides, was erected a bronze statue by the sculptor Osip Zadkine: «the Messenger» represents a feminine figure carrying the «Ship of Paris» in its arms.

– The mansion built by Comte Alexandre Waleski, the son of Napoleon I and Marie Waleswka, Napoleon III's Foreign Secretary, is on the corner of the quay and the Avenue de La Tour-Maubourg.

– The building at n°53 Quai d'Orsay is that of the SEITA. On the rather rigid-looking façade, on the third floor, two sculpted figures represent tobacco picking, whereas on the large flat pediment which is above the doorway, a large high relief represents the packing and the transport of tobacco leaves. Behind this building the Museum of the Gallery of the SEITA was opened in 1980. It is devoted to the history of the tobacco industry in France and throughout the world. This museum frequently organizes temporary exhibitions[2].

The building at n°53, on the corner of the Rue Sully-Prudhomme, was created by the architect Louis Boileau. Erected in 1933, it was richly decorated with high reliefs by the sculptor Léon Binet who worked on the prestige Parisian buildings, especially on the façades of the Hôtel Lutétia.

Sculpted Allegory by Osip Zadkine.

In the 1930s, the architect Carrol Greenough erected the flamboyant, neo-Gothic American Church and presbytery on the site of the former «Gros Caillou» (Big Pebble) tobacco factory. Gros Caillou is the name of this quarter, situated between the Invalides and the Champ de Mars.

N°67, is an Art-Deco style building. The façade in pink travertine and the angular semicircular balconies enhance its beauty. On either side of the front gate there are allegorical sculptures which represent the Seine.

The very restrained style of the building at n°89 is characteristic of the modern movement of the 1920s. It was built in 1929 by one of its most representative artists, Michel Roux-Spitz. According to an inscription, the writer and playwright Jean Giraudoux (1882-1944) lived there.

The imposing building situated on the corner of the Quai d'Orsay, the Avenue Bosquet and the Rue Cognacq-Jay is certainly one of the most peculiar ones in Paris with the entire height of its façades sculpted with stonework geometrical shapes.

Jean Nicot and tobacco

Jean Nicot (1530-1600) was born in Nîmes and died in Paris. A scholar and a diplomat, he was the French Ambassador in Lisbon in 1550 when the Portuguese presented him with a plant from the West Indies, more precisely from Cuba - tobacco - some seedlings of which were sent to Queen Catherine de Médicis, at that time Regent of France. He insisted on the therapeutic and antiseptic properties of this plant called «Pétum» and later «Tobacco». In 1674, more than half a century later, Colbert established the monopoly of development and sale of tobacco for the benefit of the state.

The Palais des Nations
on the Quai d'Orsay.
1900 World Exhibition.

The dôme of the Invalides.

The «Musée des Egouts de Paris»[3] is housed in a large underground gallery under the Place de la Résistance, facing the Pont de l'Alma.

There, in display cabinets, are a lot of objects and machinery used in the past and present for the construction and the maintenance of the sewage system, models and drawings, a slide show about the history of the sewers in Paris and, of course, the bust of the engineer Belgrand, a collaborator of the Prefet Haussmann. They were both at the origin of the system which today stretches out over 2,100 kilometres .

1. Museum open every day from 10am to 5pm. Tel: 45 55 37 67. Entrance through the Main Courtyard.

2. Museum open every day except on Sundays from 11am to 6pm. Entrance 12 Rue de Surcouf. Tel: 45 56 60 17.

3. Museum open from 11am to 5pm on Saturdays, Sundays, Mondays and Tuesdays, closed on Thursdays and Fridays. Tel: 45 05 10 29.

Living on the Seine

The first inhabited boats only appeared in Paris in the years 1965-1970. They numbered about one hundred in 1993, in general very large boats and above all old barges transformed into lodgings. Their weight must be sufficient to be able to ride the swell caused by small rapid boats, in particular motorboats and patrol boats of the river police.

The sites, in single file along the banks or in double file are granted by the management of the «Port Autonome de Paris». The mooring fees vary with the length of the boat and can reach up to 2,000 F a month. The owners and tenants must also pay a land-tax and the inhabited house duty. A barge with an engine in good condition costs about 250-300,000 francs. Its fitting out, which gives a living space of 150-200 m^2, requires an investment of about 1.5 or 2 million francs, even more in exceptional cases. The boats are only saleable with the cession of the mooring dues which can always be reconsidered by the controlling authorities. The boat-houses which moreover require constant and expensive maintenance are not therefore subject to the laws of real estate market and speculation.

In Paris, the one hundred and ten sites granted at present by the Port Autonome, are divided into several groups. From upstream to downstream, there are houseboats along the banks of the quays Saint Bernard, de la Tournelle, Montébello and de Conti. The largest floating village is that of the Port des Champs Elysées, established by the Touring-Club de France, which goes from the former Pont de Solférino to the Pont de l'Alma.

Further on, there are residents on the banks from the Quais de New York and Kennedy to the Pont de Bir-Hakeim and on the Port Suffren on the left bank, facing the Ile des Cygnes.

Living on the Seine presents some constraints but is considered as a privilege owing to the exceptional quality of the views from the portholes of every boat.

The houseboats, thanks to their presence, contribute to create a climate of friendliness and ensure security along the quays. They also form a museum of river-boats. From now on, they are a part of the landscape and are an element of the Parisian river heritage in the same way as the bridges, quays and monuments.

Their presence cannot be dissociated from the banks and everyone expects to find them there; this presence is included as part of the monumental patrimony of the capital, in the UNESCO Inventory Report, with the Classification of the Seine banks among the great world sites.

1900 World Exhibition on the Quai d'Orsay.

Construction of the Pont Alexandre III in 1898-99.

The pont Alexandre III

The Pont Alexandre III forms the monumental transversal entrance leading to the Champs Elysées.

As early as the beginning of the XIXth century a bridge had been forecast in the line of the Invalides. It was only erected between 1898 and 1900 in anticipation of the 1900 World Exhibition.

The engineers Résal and Alby designed a work with a single steel arch, spanning 115 metres and 45 metres wide. The architects Cassien-Bernard and Cousin were entrusted with the decorative part of the work for which several top level sculptors intervened.

The steel framework has fifteen equally spaced trusses. The arches are in moulded cast-iron, the superstructure in rolled steel and the decorative parts in cast-iron. This bridge was one of the first «prefabricated» works in the world, the components, especially the curves, were cast at Le Creusot, transported to the building site by barge, then installed with a spectacular travelling crane spanning the river. It only took two hundred days to construct the arch proper.

With its grandiose sculpted decoration and its elegant, bold profile, it is the most beautiful bridge in Paris. The

The Pont Alexandre III: the pillars and the equestrian allegories, the Invalides and the Esplanade.

The pont Alexandre III.

four corner pillars, 17 metres high, each of them decorated with four composite capital columns, carry gilted bronze equestrian groups on their entablatures.

On the right bank, the right-hand pillar, when facing the Invalides, supports the «Fame of Sciences» by Emmanuel Frémiet with, at its base, «Contemporary France» by Gustave Michel. The left-hand pillar supports the «Fame of Arts» by Frémiet with, at its base, «Charlemagne's France» by Alfred Lenoir. The big stone lions are by Gardet.

On the left bank, the right-hand pillar supports the «Fame of Commerce» by Pierre Granet with, at its base, «Renaissance France» by Jules Coutant. The left-hand pillar supports the «Fame of Industry» by Clément Stei-

Detail of the allegorical sculptures.

ner with, at its base, «Louis XIV's France» by Laurent Marqueste. On this bank, the stone lions are by Jules Dalou. The bank arches are profusely decorated with ornate motifs representing the sea flora and fauna.

On the keystones of the arches, upstream and downstream, two large hammer-wrought copper compositions have been placed. Upstream, they are the «Nymphs of the Seine bearing the coat of arms of Paris», downstream the «Nymphs of the Neva bearing the coat of arms of Russia». These two motifs are by the sculptor Georges Récipon, creator of the Quadriga of the Grand Palais.

An allegory on the Pont Alexandre III.

The Pont Alexandre III and the corner of the Grand Palais on the Cours la Reine side.

The pont des Invalides

In 1824, the engineer Navier set up, on this site, a single-arch suspension bridge which was not used because it did not meet the necessary stability standards. It was replaced in 1828 by another suspended work called «The Allée d'Antin» because it was not in the centre line of this avenue which led to the Rond-Point of the Champs Elysées (Avenue Franklin-Roosevelt).

This work had three spans with a clear span of 25 metres for the side arches and 68 metres for the central navigation arch. In anticipation of the World Exhibition in 1855 which was due to take place on the Champs Elysées, at the Palais de l'Industrie et des Beaux Arts and in the Galerie des Machines on the Cours La Reine, replacement of the suspension bridge by a larger masonry work, called the «Pont des Invalides», was undertaken in 1854. It was built in less than a year.

It had four depressed semicircular arches with a clear span of 32 metres, supported on five-metre thick piers; the deck was sixteen metres wide. The engineers were de Lagallisserie and Vaudrey.

In 1879-1880 as a result of damage caused by the breaking-up of the ice on the Seine, the bridge was almost entirely rebuilt, but without any architectural modifications. It kept its bossed stonework and its moulded cast-iron balustrades. On the pierheads of the central pier, the sculpted allegories were put back into place: the «Victoire Maritime» (Victory at Sea) by Victor Vilain and the «Victoire Terrestre» (Victory on Land) by Georges Diébolt. On the tympana above the left and right lateral piers, large escutcheons with men's heads surrounded by clustered flags were sculpted by Lavergne and Horace Daillon.

A sculpted allegory on the Pont des Invalides: the «Victoire Maritime».

The pont de l'Alma

Only the famous «Zouave» is left of the old Pont de l'Alma, built in 1854 and named after the victory in Crimea.

The work, completed, as many others, during Haussman's town replanning, was intended to link the new quarters which were built on the left bank, round the new Avenues Rapp and Bosquet, and those which extended on the right bank, on the Chaillot hill, from the Avenue Montaigne to the Etoile and the Trocadéro.

Like the Pont des Invalides, the works were planned to be completed for the opening of the World Exhibition in the Spring of 1855.

It had three elliptical arches, spanning 43.4 metres in the centre and 39 metres on the banks. These arches rested on 5-metre thick piers and 8-metre thick abutments. It was 20 metres wide.

On the pierheads, the statues of four soldiers were placed. Only the Zouave by Diébolt was put back in place when the new bridge was built. The «Chasseur» by Arnaud is near the redoubt of Vincennes-Saint-Maurice, under a flyover of the A4 motorway. The «Artilleur» by the same sculptor is in La Fère. The «Grenadier» by Diébolt is in Dijon (these four forces took part in the battle of the Alma).

The engineers of the Pont de l'Alma were de Lagallisserie, Darcel and Vaudrey.

The Pont de l'Alma and the Colline de Chaillot in 1866. A painting by Stanislas Lépine.

The pont de l'Alma in 1890.

The new bridge, completed in 1972, is a steel work resting on a single pier situated towards the right bank. Built by the «Direction des Services de la Navigation», with the architect Auguste Arsac, the present Alma bridge is 40 metres wide and 136 metres long.

The Zouave of the Alma Bridge.

The passerelle Debilly

This «temporary» foot-bridge was built in 1900 to facilitate visiting the palaces of the World Exhibition, at that time, on the banks of the Seine. It is an elegant construction with a riveted steel framework. The two tranverse bracing arches are supported by masonry piers near the banks.

It commemorates the memory of General de Billy, killed at the battle of Iéna and whose name was given to the right bank quay on the order of Napoleon (today Quai de New York). It faces the museum of Modern Art.

The Passerelle Debilly.

From horse-drawn passenger barges to passenger riverboats.

Several navigation companies offer tourist cruises on the Seine and the Canal Saint-Martin, the loop of the Marne: the Bateaux Mouches, the Vedettes du Pont Neuf, the Bateaux Parisiens, the Vedettes Paris Ile-de-France, the Croisières Paris-Canal and the Canauxrama Company.

For centuries, the Seine has had heavy passenger traffic.

Before the Second Empire, this traffic was essentially upstream, from the Ports Saint-Paul and des Celestins alongside which moored the canal-boats and passenger barges bound for Bourgogne, Champagne, the centre of France and even Nantes by means of the Loing canal and the Loire or for Lyons and Marseilles by the Saône and the Rhône.

After the canalization of the river, the dredging and the construction of large gauge locks, a gigantic enterprise began under Louis Philippe and Napoleon III. River and sea links were created for passengers between the Port du Louvre and... London, by the Seine and through Rouen by means of small sea-going steamers which sailed until the 1930s.

Short distance passenger transport in Paris and the near suburbs, from

The Pont Alexandre III at the time of the 1900 World Exhibition.

An omnibus boat near the Alma, circa 1900.

Charenton to Suresnes, grew the most during the Belle Epoque from 1880 to 1914.

The omnibus boats at that time carried twenty-six million people a year, on average, that is to say seventy thousand passengers a day but with very important «peak» days on Sundays when the city people went in great numbers to Meudon, Saint Cloud, Suresnes and the racecourse at Longchamp.

The omnibus boats ceased sailing in 1934 because of the competition from trams, buses, the underground railway and above all cars.

Downstream, between the Pont Royal and the Pont de Sèvres, from the end of the XVIIth century to the beginning of the XIXth century, they sailed regularly, especially on days when the fountains were flowing at Versailles and Saint Cloud.

These horse-drawn boats were replaced by steamers. The first test was carried out on the Quai de New York, at that time Quai des Bonshommes, by the engineer Robert Fulton on August 9th 1803. The Marquis de Jouffroy d'Abbans experimented further in 1816 and 1817 on a boat called the «Génie du Commerce».

These prototypes generated a flotilla of paddle steamers which developed considerably throughout the XIXth century.

As early as 1837, the «Compagnie Générale des Bateaux à Vapeur» provided a service for the daily traffic from Paris to Saint Cloud, connecting with the bus lines.

The navigation companies were reorganized in 1867 on the occasion of the Second Parisian World Exhibition which took place on the Champ de Mars. The paddleboats were progressively replaced by propellor-driven boats.

A new company took the place of the previous ones: the Compagnie des Bateaux Omnibus. In 1878, at the time of the Third World Exhibition, it merged with its competitor, the Compagnie des Hirondelles Parisiennes. They were replaced in 1886 by the Compagnie Générale des Bateaux Parisiens which, from that time on, ran one hundred and five boats, one hundred landing stages and an installation of maintenance and boat sheds near the Quai du Point du Jour.

Traffic was maintained between Maisons-Alfort and Suresnes until the disappearance of the omnibus boats in 1934.

It was only in the fifties that the first company appeared, organizing daily tourist cruises on the Seine: The Compagnie des Bateaux Mouches which used restored omnibus boats for a few years. At that time it had a prestige boat, the very elegant «Borde Fré-

The port de Paris

The Port de Paris was, in the XIXth century, the most important port in France. In 1824, 15,276 boats and 4,500 timber rafts stopped there. From that date on, a lot of boats passed via the docks of the Villette, the canals Saint-Martin, of the Ourcq and Saint-Denis.

The Seine, Paris' vital artery, was often a blocked artery: in Summer, when the waters were low, the boats could only sail with a third or a quarter of their normal load.

In Winter, navigation was sometimes interrupted because the Seine was frozen over whereas the melting of the snow raised the water level which prevented boats from passing under the bridges.

All this led to a huge congestion of the quays, banks and unloading sites. This situation improved progressively throughout the XIXth century and especially from 1850 onwards with the dredging of the river, the installation of new locks to ensure a constant water-level, the construction of quays, the improvement of the low banks and also the decrease of river traffic after the proliferation of the railways converging on the capital.

The canalization of the Seine was entirely finished at the end of the XIXth century, which caused the inversion of the traffic evolution and of the development of Paris and its suburbs to the West.

As a result, from that time on, the traffic from upstream declined almost to the point of disappearing. Traffic coming from the low Seine went on increasing from the moment Rouen and Le Havre became the outer ports of Paris. In 1909, nine million tonnes of goods reached the ports of Paris on the Seine and la Villette, mainly from Rouen.

New important economic activity settled on the banks of the river and spread from Grenelle to Gennevilliers.

Today, the Port Autonome de Paris concentrates most of its installations in the West of the capital, more particularly at Gennevilliers. In the East, the Port de Bonneuil is large and diversified.

In 1992, the volume of goods processed at the installations of the Port Autonome de Paris represented more than twenty-seven million tonnes, a big part of which were heavy materials (sand, gravel and coal) and oil products.

Port Autonome de Paris: 2 quai de Grenelle, 75015 Paris. Tel: 40 58 29 99.

An omnibus boat along the landing-stage of the Point du Jour at Auteuil.

The Bateaux Parisiens in front of the Palais du Champ de Mars at the time of the 1900 World Exhibition.

tigny» on which, in 1956, Queen Elizabeth of England and the President of the Republic René Coty made an official visit of the Quays of Paris.

Today the «Bateaux Mouches» have a dozen boats, each with several hundred seats and with double panoramic decks. Some of these boats offer lunch or dinner on board during the cruise.

Navigation on the Seine in 1900.

The Bateaux Parisiens, created in the 1960s also have vessels on which various cruises are organized, among them dinner cruises. It is the same for the Vedettes du Pont Neuf and the Vedettes Paris Ile-de-France. The Croisières Paris Canal also propose lunches on board or boats equipped for seminars. Batobus, set up in Paris in 1989, provide regular links between their five landing stages spread out along the banks, between the Port de La Bourdonnais and the Hôtel de Ville.

In total, the tourist boats of the different companies carry 6.6 million passengers a year (1992). About 70 per cent of these passengers are foreign tourists.

- Bateaux Parisiens Notre Dame, Quai de Montebello, Saint-Michel, 75005. Tel: 43 26 92 55.
- Bateaux Mouches, Berge du Cours la Reine, Pont de l'Alma, 75008. Tel: 42 25 96 10 / 40 76 99 99 / Fax: 42 25 02 28.
- Bateaux Parisiens Tour Eiffel, Port de La Bourdonnais, 75008. Tel: 44 11 33 44.
- Vedettes du Pont Neuf, Square du Vert Galant 75001. Tel: 46 33 98 38.
- Vedettes Paris Ile-de-France, Port Suffren, Pont d'Iéna on the left bank, 75007. Tel: 47 05 71 29.
- Canauxrama, Bassin de la Villette, Quai de La Loire, 75019. Tel: 42 39 15 00 / Fax: 42 39 11 24.
- Paris Canal, 19-21, Quai de La Loire, Bassin de la Villette, 75019. Tel: 42 40 96 97 / Fax: 42 40 77 30.

The pont d'Iéna

This bridge, which was to be called «Pont du Champ de Mars», was planned in 1806. It was given the name of Iéna because on 14th October of that year Napoleon won the battle of Iéna in Germany against the Russians and the Austrians.

The work designed by the engineer Lamandé, completed in 1814, was intended to join the Champ de Mars to the new quarter Napoleon wanted to build on the Chaillot hill around the Palais du Roi de Rome.

The Pont d'Iéna was originally 14 metres wide. Its width was doubled for the 1900 and 1937 World Exhibitions, while keeping the same profile and reusing the ornamental sculpture of the tympana and pylons. It has five depressed arches with a clear span of 28 metres resting on 3-metre thick piers.

The tympana, above the semicircular pierheads were decorated with big eagles in high relief surrounded by laurel crowns, by the sculptor François Lemot.

At the entrances of the bridge, four equestrian groups, sculpted in the round, were placed on monumental pylons at the beginning of the Second Empire. These allegories represent a Guerrier Gaulois (Gallic Warrior) by Antoine Préault and a Guerrier Romain (Roman Warrior) by Louis Daumas, on the right bank and a Guerrier Arabe (Arab Warrior) by J.H. Feuchère and a Guerrier Grec (Greek Warrior) by François Delvaux on the left bank.

The widening of the Pont d'Iéna in 1910 facing the Palais du Trocadéro and its gardens.

The quai de New York

Begins at the Pont de l'Alma
and ends at Rue Beethoven.

This quay used to be called «des Bonshommes» because of the presence at Chaillot of the convent of the same name and later, in 1610, «La Savonnerie» because the carpet factory settled there. It then became the Quai Chaillot and later, in 1807, the Quai de Billy, in honour of General de Billy killed at Iéna. In 1918, it was named «Quai de Tokyo» and finally in 1945 it became the Quai de New York.

At the end of the quay, near the Place de l'Alma, a gold-leaf flame, a reduced-scale copy of the flame of the Statue of Liberty, was installed on a bronze column. It was a by the sculptor Bartholdi. This monument was erected by the International Herald Tribune on the occasion of its centenary.

The Quai de New York in 1900. A district of Old Paris reconstituted on the bank.

The Rue des Frères Perrier and the Rue Debrousse are on the site of the old «Chaillot» fire pump.

The Museum of Modern Art was built in 1937 by the architects Dondel, Aubert, Viard and Dastuque on a site on which the «La Savonnerie» carpet factory stood in the XVIIIth century.

The two wings of the building are set around a high central peristyle with cylindrical columns allowing a view towards the Palace Galiera. The building has luxurious by carved decorations. The grand staircase that leads to the upper peristyle is surmounted by two huge low reliefs by Alfred Janniot, one of the best sculptors of the 1930s.

On the wall which delimits the upper terrace, are three works symbolizing France, Strength and Victory by Bourdelle.

Two museums there are in the symmetrical wings of this large building overlooking the quays. The National Museum of Modern Art, most of whose collections are at Beaubourg, was created here. The Museum of Modern Art of the City of Paris, situated in the east wing, on the right, contains the collections of the City of Paris. It periodically organizes great thematic retrospectives or exhibitions devoted to an artist or to a trend of Modern and Contemporary Art[1].

A few neo-Classical buildings, ancient town houses, one of which houses the Noma Bismarck Foundation, separate the Museum of Modern Art from the vast gardens of the Trocadero.

Napoleon I planned to build a gigantic palace on the Chaillot hill for his son, the King of Rome. The architects, Percier and Fontaine, only had time to lay its foundations. In 1860, during Haussman's great town replanning, the hill was levelled, the Place du Trocadéro was created and radiating avenues were opened. A vast terrace was laid out with a view over the Champ de Mars.

Gardens were laid out down to the quay. A few years later, Gabriel Davioud and the engineer Bourdais built the Palace of the 1878 World Exhibition which remained in the Parisian landscape until 1935. It was then replaced by the present

The Museum of Modern art.

Sculptures and colonnades.

Photography centre: open every day except on Tuesdays from 9.45am to 5pm.
Tel: 47 23 36 53.

1. Modern Art Museum of the City of Paris: open every day from 10am to 5.30pm (8.30 pm on Wednesdays); closed Mondays.
Tel: 47 23 61 27 / 40 70 11 10.

building, a work by the architects Carlu, Boileau and Azéma.

The Palais de Chaillot marks the landscape of the banks of the Seine in a particularly impressive way. The brilliant idea consisted of creating a huge opening, a monumental portico, defined by two large symmetrical curved wings on top of the hill. This idea was an innovation, a daring gesture forming an imposing void between the two enveloping wings of the palace. It was also a simple idea which consisted of laying out, in this privileged place, a site from which the beauty of the whole city could be seen. This open-air theatre has since been used many times for festivals, feasts and celebrations.

The works, in the gardens which slope down gently towards the quay, were carried out for the 1937 World Exhibition.

The large central ornamental lake on which water canons spout, is accompanied on the left by a sculpture representing the «Joy of Living» by Leon Driver and on the right by «Youth» by Pierre Poisson.

The Jardins de Chaillot are on the old site of a country house built in 1583 for Queen Catherine de Médicis. This house had terraces stretching

Sculptures of the thirties

Here, the sculptures are equally abundant in the gardens and on the walls. This anthology took its inspiration from various sources; Antiquity, the Middle Ages, Exotism and Cubism. One of the merits of the Palais de Chaillot is to have made space for monumental art and today to exhibit the most complete set of the figurative trends of the 1930s, with well-known sculptors such as Henri Bouchard, Paul Belmondo, Paul Niclausse, Raymond Delamare, Paul Landowski, Georges Saupique, Hubert Yencesse, Jacques Zwoboda, etc.

The Palais du Trocadero erected for the 1878 World Exhibition. A lithography.

The palais de Chaillot in 1937.

down to the river. The Maréchal de Bassompierre, a friend and a confident of Henri IV, bought this house. He left it for a twelve year stay in the Bastille having apparently burnt six hundred love letters...

Musée de l'Homme: open every day except on Tuesdays from 9.45am to 5.15pm. Tel: 44 05 72 72.

Musée de la Marine: open every day except on Tuesdays from 10am to 6pm. Tel: 45 53 31 70.

Musée des Monuments Français: open every day except on Tuesdays from 10.30am to 5pm. Tel: 44 05 39 10.

Musée du Cinéma and Cinémathèque: open every day except on Mondays from 10am to 5pm.

The museums of Chaillot

The Palais de Chaillot has several Museums(*): to the west of the left wing, in the direction of Passy, the Musée de la Marine (Maritime Museum) and the Musée de l'Homme (Museum of Mankind); to the east of the right wing, the extraordinary Musée de la Sculpture Comparée (Comparative Sculpture Museum) or the Musée des Monuments Français (Museum of French Monuments), installed here in 1879 on the initiative of Viollet-Leduc. A Musée du Cinéma (Cinema Museum) and a Cinémathèque (Film Library) were set up in this wing on the initiative of Henri Langlois, a cinematic historian. The Théâtre National Populaire (TNP), created by Jean Vilar, still occupies the large theatre situated in the basement beneath the upper terrace.

The Fountains of the Jardins de Chaillot.

The quai Edouard-Branly

Begins at the Pont de l'Alma
and ends at the Pont de Bir-Hakeim.

This quay was named in 1941 after the physician Edouard Branly (1844-1940). It was part of the Quai d'Orsay. The high bank forms a long, pleasant walk along the Seine, from which there is a panoramic view of Chaillot and Passy.

A huge building with a monumental doorway stands at n°11. It was built under the Second Empire by the architect Lefuel, for the stables of Napoleon III. Today it is the garage of the Presidency of the Republic.

High buildings were erected at the end of last century on each side of the esplanade of the Champ de Mars where the Eiffel Tower has stood since 1889. Having been, since the beginning of the century, a research laboratory for the transmission of radio and then of television waves, the tower still serves as a transmitter. It remains the most visited monument in Paris. The bust of Gus-

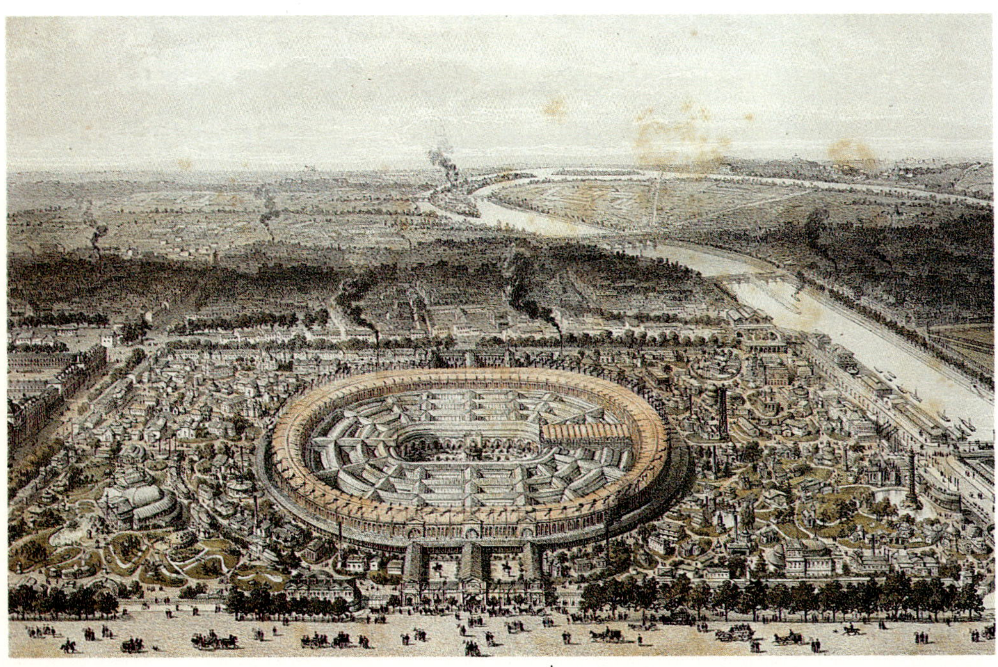

The 1867 World Exhibition on the Champ de Mars.

tave Eiffel, at the foot of one of the pillars, is by Antoine Bourdelle[1].

With the construction of the Ecole Militaire (French Military Academy) under Louis XV, by the architect Jacques-Ange Gabriel, the part of the Plaine de Grenelle that stretched out to the Seine was converted into «the Champ de Mars» that is to say a training ground for the Cadets of this Military Academy. During the Revolution, it was the political meeting place of hundreds of thousands of people for the «Fête de la Fédération» in which Talleyrand took part on July 14th 1790.

In the XIXth century, many festivities were organized there. It was, for half a century, the racecourse of Paris before the opening of the Longchamp hippodrome in 1855. On the occasion of the 1867, 1878, 1889, 1900 World Exhibitions, it was covered with palaces which were as splendid as they were transient.

The architect Jacques-Ange Gabriel decided to have the buildings of the Ecole Militaire facing the Seine and to construct the main façade on that side because of the extremely monumental effect it would have, the interest of which he had realized when imagining the esplanade he wished to lay out between that building and the river.

The Palais of the Champ de Mars, in 1900.

The illuminated Eiffel Tower. 1889 World Exhibition.

The Eiffel Tower

The engineer Gustave Eiffel completed this tower in just 26 months. The construction caused much controversy. Only two hundred workers worked on the building site, a feat made possible thanks to the methodical preparation of pieces ready to be fitted together in the workshops of Levallois. The tower is 318 metres high and weighs 7,500 tonnes. Its base area is a 130-metre sided square.

At the end of the terrace that overlooks the low banks and contains the Paris-Versailles (via Meudon) railway and the C line of the RER, two small monuments were erected slightly upstream of the Pont de Bir-Hakeim. One honours the memory of General Diégo Brosset (1898-1944) a hero of the Resistance and FFI action («Forces Françaises de l'Intérieur») and was achieved following the plans of the architect Béatrice Delamare. The marble bust was sculpted by Raymond Delamare. In 1989, the bust of the explorer Jean-François de Galaup was placed nearby. He was the Comte de la Pérouse (1741-1788) who disappeared in the Pacific on the island of Vanikoro. This daring sailor was charged by Louis XVI, in 1785, with finding a passage North-West of America which had been identified by Cook and Clarke, and with reconnoitring the waters off Japan and the Solomon Islands. This bronze sculpture by O. Dabio was offered to the Ville de Paris in 1989 by the nearby Australian Embassy.

The «Chateau» (Central Pavilion), designed by Gabriel has, on the Seine side, a large sally with Corinthian columns flanked by two wings. It is surmounted by a quadrangular dome itself resting on a high attic behind the triangular pediment.

From 1806 onwards, with the construction of the Pont d'Iéna, the planting out of the Champ de Mars and the linking of the Ile des Cygnes with the left bank, the perspective which Gabriel had wanted in front of the «Château, finally came true.

Under the Second Empire, the laying out of the viewpoint on Chaillot Hill, then the construction of the Palais du Trocadéro in 1878 completed this composition, whose centre line led to the large rotunda of this palace, which was pulled down in 1936. In 1937, the architect Carlu added the finishing touches by creating the terrace, in the form of an esplanade between the two wings of the new palace. From then on there existed a clear space from the left bank to the right bank, from the Ecole Militaire to the Trocadéro, with only the Eiffel Tower in the middle.

1. Visit every day: levels 1 and 2 from 9.30am to 11pm; level 3 from 9am to 10.30pm. Tel: 45 50 34 56.

The Champ de Mars and the Eiffel Tower seen from the Palais de Chaillot.

Following Page: *Illuminations of the Eiffel Tower.*

The ancient pont de Passy and the pont de Bir-Hakeim

A foot-bridge was built on this site for the 1878 World Exhibition. The platform of the Ile des Cygnes separated it into two parts. On the big arm on the right bank, the foot-bridge was 120 metres long, on the small arm 90 metres.

This bridge, essentially used for pedestrians, was moved in 1904 at the time of the construction of the elevated railway for the Metro, and placed in the line of the Boulevard de Grenelle on the left bank and of the Rue de l'Alboni on the right bank. Its width

The Passerelle de Passy in 1880. A water-colour by Fraipont.

Detail of the sculptures of the Pont de Bir-Hakeim.

Equestrian allegory of Renaissance France on the upstream tip of the Ile des Cygnes.

The bridge and the buildings on the Quai Kennedy.

was doubled and the viaduct of the Metropolitain was built in the central line of the roadway.

On the piers of the bridge were placed eight monumental cast-iron groups by the sculptor Gustave Michel. They represent blacksmiths and boatmen.

The viaduct of the Metro, which rests in the centre on a stone pillar, was built by the engineer Louis Bette, in association with the architect Jules Formigé.

The double deck bridge rests on the north tip of the Ile des Cygnes. On the platform stands a statue dedicated to «Renaissance France» by the Danish sculptor Wederkink (1930).

The viaduct bridge of Bir-Hakeim, 257 metres long and 25 metres wide, was named in 1949 in commemoration of the victory of General Koenig against Rommel's army at Bir-Hakeim in Libya in June 1942.

Restored in 1933, this work has been reequipped with the streetlamps designed at the beginning of the century and which disappeared in the fifties.

The Pont de Bir-Hakeim.

The quai de Grenelle and the Ile des Cygnes

Begins at the Pont de Bir-Hakeim and ends at the Pont de Grenelle.

The quay is separated from the Seine and the banks by the railway running from les Invalides to Versailles and the RER lines.

The Ile des Cygnes stretches out opposite the Quai de Grenelle from the Pont de Bir-Hakeim to the Pont de Grenelle interrupted by the bridge of the former Railway Passy-Invalides, today incorporated into the C line of the RER.

850 metres long and 11 metres wide on the upper part, this tree-lined walk was created in 1825-30 to protect the Port de Grenelle, installed by the property developers from the «Beau Grenelle» district.

It was called the «Ile des Cygnes» in memory of the island that occupied a long strip of land between the Pont des Invalides and the present Pont d'Iéna, which was joined to the left bank at the time of the laying out of the Quai d'Orsay opposite the village of Gros Caillou.

The Quai de Grenelle under the snow, circa 1890. In the background, the Palais du Trocadéro. A painting by Galien Laloue.

The railway bridge (today RER) crossing the Ile des Cygnes.

Swans, sent to Louis XIV in 1676 by the Kings of Denmark and Sweden, were placed on this island which was previously called «Ile Maquerelle», a distortion of male-quarrel. This deserted island was used as a meeting-place by duellists.

Before the construction of the buildings of the «Front de Seine» begun in the sixties, which was at the time occupied by industries, there was a famous «guinguette» (open-air café), «le Bal de la Marine», near the Pont de Grenelle.

The Direction du Port Autonome de Paris, a public body that controls all the harbour activities of the Ile-de-France Region, is installed on the bank of the Quai de Grenelle near the Pont de Bir-Hakeim.

Contemporary buildings along the Quai de Grenelle.

The quai J.F.-Kennedy

Begins Rue Beethoven
and ends at the Pont de Grenelle.

This quay is an ancient section of the road from Paris to Versailles. The «des Bonshommes» barrier, as it was called, marked the limits of the city of Paris, level with the present Pont de Bir-Hakeim. It was a construction by the architect Claude-Nicolas Ledoux for the Fermiers Généraux.

Jean-Jacques Rousseau, who used to use the spa at Passy, lived in a house on the quay in 1750. The Château de Passy and the Hôtel de Lamballe dominated the panorama. The gardens sloped down to the road to Versailles. In 1801, Benjamin Delessert established the first sugar-beet refinery in the park of his country-house. The Delessert Park stretched out from the quay to the present Rue Raynouard. The «Parc de Passy» was occupied for fifty years by the «temporary» buildings of the Ministry of the Environment... Now there is talk of laying out a large public garden here.

The Maison de la Radio (Radio-France House) built in the years 1958-62 by the architect Henri Bernard is on the site of the ancient Passy gas-

Passy, a spa

The village of Passy used to be a popular spa. The narrow, picturesque Rue des Eaux, starting at n°18 Quai de Passy and climbing up to the Rue Raynouard, was named in memory of this spa.
This lane was opened around 1650 on the occasion of the discovery of the first ferruginous springs which were successful up to the time of Jean-Jacques Rousseau.

The banks of Passy and Auteuil in the middle of the XVIIIth century. A painting by Raguenet, Carnavalet.

The parks of Passy and the road to Versailles in the XVIIIth century. A painting by Gravenbroek, Carnavalet.

works which had itself replaced the gardens at the end of the XIXth century[1].

Since 1964 the former quai de Passy has honoured the memory of President J.F Kennedy.

---

1. Radio-France House, 116 Av. J.F Kennedy. Tel: 42 30 21 80. Guided tours every day except on Sundays and holidays: 10.30am, 11.30am, 2.30pm, 3.30pm, 4.30pm.

Opposite the Quai Kennedy, near the Passy Métro station: The Caveau des Echansons, Wine Museum, Rue des Eaux. Tel: 45 25 63 26. Open every day from 12am to 6pm.

The statue of Liberty at the tip of the Ile des Cygnes and the Maison de la Radio.

The pont de Grenelle

In the years 1825-1830, while he was developing the Beaugrenelle quarter, the builder Violet was allowed to install a port and a landing stage on the small arm of the Seine on the left bank, at right angles with the Ile des Cygnes, as well as a bridge linking the village of Passy to the road to Versailles.

This wooden bridge had six arches. It was entirely rebuilt in 1875 by the engineers Vaudrey and Pesson and the Ateliers Cail. The 10-metre wide deck rested on a series of arcatures on cast-iron arches which spanned 25 metres.

This bridge was pulled down in 1966 and replaced by a new work which, supported by the earth platform of the Ile des Cygnes, crosses the two arms of the river with a single arch on each side.

A scale model of the statue by Auguste Bartholdi, «Liberty Lighting the World» was placed at the downstream tip of the Ile des Cygnes in 1885.

This monumental sculpture was restored in 1990 and given a bronze coloured patina.

The Ile des Cygnes, circa 1900.
A water-colour by Leverd.

The pont Mirabeau

The Pont Mirabeau is, like the Pont Alexandre III, a technical masterpiece and an example of architectural elegance. It was designed by the engineer Résal and achieved with the help of the engineers Alby and Rabel. 20 metres wide, this work rests on piers whose foundations are 16 metres under the average level of the river. The piers are covered with Cherbourg granite and limestone. The central arch has a clear span of 93.2 metres; the two lateral arches 32.4 metres.

Each arch has seven rolled steel trusses resting on piers by means of ball-joints. The two central half-trusses are hinged at the key.

The cornices and guard-rails are in moulded cast-iron as are the vertical posts of the bank arches. The escutcheons in the centre of the bridge are in bronze. The semicircular pierheads are decorated with four bronze sculptures by Antoine Insalbert. These are allegories representing sea divinities. The work was inaugurated in 1898.

A plaque, placed in 1986 on the occasion of its classification in the Inventaire des Monuments Historiques, tells that the Pont Mirabeau was the source of inspiration for the very famous poem by Guillaume Apollinaire (1880-1918)...

The Pont Mirabeau. Details of Sculptures.

The quai André-Citroën

Begins at the Pont de Grenelle
and ends at the Pont de Garigliano.

The Eau de Javel

The old village of Javel dates back to the XVth century. In the XVIIth century its windmill was well-known to the Parisians. The bathers and the boatmen used to come and relax in these parts, which were still bucolic. In 1777, various manufacturers received from the Comte d'Artois, who controlled the villages of Javel, Grenelle and Issy, the right to create factories to produce potassium hypochlorite which was called «Eau de Javel» (bleach). Throughout the XIXth century, other factories flourished along the quay especially the industries linked with iron work and ship repairs.

This quay is on the site of an ancient towpath. It was called Quai de Javel in 1843, and then Quai André Citroën in 1958 in homage to the car maker who had set up his main factory there in the 1920s.

The Port of Javel was installed on the banks below the Quai André Citroën, in 1866. There are presently different concessions belonging to the Pont Autonome de Paris, especially installations for the storing of sand and gravel and concrete mixing plants. These industries will probably disappear in the short term to make space for a promenade along the banks; a logical extension of the new André Citroën Park.

Not far from the Pont Mirabeau, the offices and studios of the TV Channel «Canal Plus» were designed by the American architect Richard Meyer. At the other end, near the rue Louis

The «Eaux de Javel» factory in the XIXth century.

André Citroën Park inaugurated in 1992.

Leblanc, the architect Olivier-Clément Cacoub erected the building «the Ponant».

André Citroën Park, opened to the public in 1992, is on the site of the former car factory. Built by the «Direction des Parcs et Jardins», on the initiative of the Ville de Paris, this park was designed by the landscape gardeners and architects Patrick Berger, Gilles Clément, Alain Provost and Jean- Paul Viguier.

With an area of about fifteen hectares, laid out around a huge central lake, this park boasts plants, groves, waterworks and waterfalls. Two large tropical greenhouses stand on each side of the main line of the perspective which will soon be extended to the Seine.

The «Canal Plus» building designed by Richard Meyer.

The quai Louis-Blériot

Begins at the Pont de Grenelle
and ends at the Pont de Garigliano

The shipyard of the Bateaux-Mouches on the Point du Jour.

An old towpath was transformed to produce this quay in 1883. It was called Quai d'Auteuil up to 1937, when it was renamed after the pilot Louis Blériot.

Most of the buildings on this quay were erected in the thirties. Some, such as the one on the corner of the Rue Degas, which is entirely covered with an Art Deco mosaïc in delicate shades, are quite outstanding. The Quai Louis Blériot has lost its banks, which have now become the «Right Bank Expressway».

The Quai Louis Blériot was followed, in the direction of Boulogne, by the Quai du Pont du Jour, today dedicated to the memory of the pilot and writer Antoine de Saint-Exupéry.

The name «Point du Jour» was formerly synonymous with «guinguettes», festivities, cabarets, fine Sundays on the waterside accompanied with accordion-tunes from the «dancing halls».

The name «Point du Jour» (day-break) may seem paradoxical for a place which is to the west of Paris. The explanation given by the ancient his-

Anglers along the Quai du Point du Jour at the beginning of the century. A painting by Ener.

The Porte de Passy and the Pont d'Iéna, circa 1820. A print by Hemely.

torians is that this name originated from a duel between the Marquis de Coigny and the Prince de Dombes. After a quarrel over a game of cards at the Court, at Versailles during the reign of Louis XV, these two gentlemen decided to meet on the road to Paris «au point du jour», which is to say when the light was strong enough to fight a duel.

Omnibus boats and ginguettes at the Point du Jour, circa 1895.

From the viaduc d'Auteuil to the pont de Garigliano

The Pont Viaduc du Pont du Jour, destroyed at the beginning of the sixties, was the most spectacular of all the bridges in Paris: its double row of archways gave the impression of a Roman aqueduct. This work marked the downstream limit of Paris .

175 metres long and 30 metres wide, this viaduct was built to allow the passing of the circular railway line created under the Second Empire. The road bridge had five elliptical arches with a clear span of 30.25 metres and a rise of 9 metres. The head room was higher than those of all the other bridges of Paris. The viaduct of the railway in the centre line of the bridge was 9 metres wide. It had a series of thirty-one semicircular arches. This viaduct went on to the end of the Boulevard Exelmans. Magnificent bossed stonework gave this work a splendid appearance. Burrstone and freestone were used in abundance.

The Pont Viaduc d'Auteuil was destroyed in 1962. The Pont de Garigliano, completed in 1966, replaced it.

Its name comes from the small Italian river near which the French troops fought in 1943-44 during the battle of Monte-Cassino. In 1502, Bayard single-handedly defended the entrance of a bridge on the same river, against the Spaniards... The work has a steel framework; the trusses of the arches rest on piers in white reinforced concrete which were built with particular care.

The ring road bridge cuts the perspective of the Seine a few hundred metres downstream of the Pont de Garigliano and marks the territorial limit of the Ville de Paris.

The Pont-Viaduc du Point du Jour in 1879. A painting by Pierre Prins.

A stroll along the banks

One of the most pleasant ways of discovering the sights of the Seine and Paris is to take a stroll along the banks. However, this stroll is only possible on part of the banks, either because the are occupied by the expressways on the right and left banks, by certain installations of the Port Autonome with sand and gravel yards, concrete mixing plants or simply because, in some places, there are no banks at all.

The Ile Saint-Louis, the Quais de Béthune and d'Orléans oriented southwest have virtually no banks, these only begin at the very end of the Quai d'Orléans near the Pont Saint-Louis and continue on the northeast side of the Ile, on the Quais Bourbon and Anjou where moreover they are rather narrow. However there are beautiful vantage points at the downstream end from where the succession of bridges on the big arm of the right bank can be seen. At the upstream end, there is a narrow walk at the foot of the wall which marks off the public garden. There is a wonderful view of the river upstream and of the Saint-Bernard gardens.

Round the Ile de la Cité, there are few low banks because of the extreme density of the buildings and houses that occupied the Ile in the past. At right angles with the Parvis Notre-Dame, on the Quai du Marché-Neuf, the bank is only a narrow passage, greatly appreciated however because it faces south. On the side of the Quai de l'Horloge, at right angles with the houses on the Place Dauphine, the bank is also very narrow. On the south side, towards the Quai des Orfèvres, it offers a walking space, all the more appreciated as it leads into the Vert-Galant public garden where there is, on the downstream side, the most beautiful view of Paris and of the succession of bridges, with on the left and on the right the palaces of the Mint, the Institute, Orsay, the Louvre and the Grand Palais in the distance.

Upstream, on the left bank, from the Pont National to the Pont d'Austerlitz, the banks are occupied by building material yards, and on the Quai d'Austerlitz by a huge warehouse which will hopefully soon be demolished.

Opposite the Grande Bibliothèque, between the Ponts de Tolbiac and de Bercy, a walk is being built. Towards Bercy, the part near the Pont National is occupied by a big concrete mixing plant. But the largest part is a long pleasant walk which will soon have several «guinguettes». The bank of the Quai de la Râpée is also to be converted back into a walk, after the construction of the Pont Charles-de-Gaulle.

On the Quai Henri IV, the bank was rearranged and cleared of warehouses in 1992 after the neighbouring expressway was passed underground. When passing under the Pont Morland and the lock of the Arsenal, the pedestrian can go into the gardens of the Bassin de l'Arsenal. Because of the expressway, the right bank is impassable between the Pont de Sully and the Louvre except for a garden accessible from the Quai de l'Hôtel-de-Ville.

Continuing downstream, there is a long sunny walk that begins near the foot-bridge of the Pont des Arts and stretches down to the Pont de l'Alma (it has however been temporarily blocked by the right bank abutment of the Pont de Solférino).

The partial reconstruction in the 1960s of the Quai de New York, downstream from the Pont de l'Alma, has removed the low bank where the walk could however be resumed if there was a wide foot-bridge overhanging the river.

Beyond the Pont de Bir-Hakeim, the expressway runs along the bank down to the Point du Jour. In the future, solutions similar to those which were used on the Quai Henri IV might be adopted to allow the creation of a high bank above the expressway.

This solution would also be possible on the left bank, at right angles to the Quai Anatole France.

On the left bank, the most beautiful walk is that which begins at the Pont d'Austerlitz, with the Jardins Saint-Bernard, and continues at right angles with the Quais de la Tournelle and Montebello to become, further down, a narrow passage on the Quai Saint-Michel and then to widen out halfway along the Quai des Grands-Augustins down to the approaches of the Pont Marie under whose arch a passage could also be planned. Near the Pont des Arts is the theatre-boat «La Mare au Diable». The left bank expressway runs along the quays between the Pont Marie and the Port de La Bourdonnais, rendering the walk impassable. This drawback is partly compensated by the long terrace, a viewpoint bordering the Quai d'Orsay after the Pont des Invalides, and the Quai Edouard-Branly to the Pont de Bir-Hakeim.

The walk on the low bank is again possible at right angles with the Ports de La Bourdonnais and de Suffren at the Pont de Bir-Hakeim.

A walk overlooking the railway lines of the RER was laid out near the Quai de Grenelle.

The Ile des Cygnes is of course, another privileged part of this pedestrian way which will soon continue at right angles with André Citroën Park from where the port facilities will progressively be cleared.

The canal Saint-Martin and La Villette

The Canal Saint-Martin was part of the system which enabled river traffic to by-pass Paris, thus avoiding the bridges. It took several centuries to achieve this project... The first study dates back to 1529.

During the reign of Louis XVI, the engineer Jean-Pierre Brulée took up the project which consisted of collecting the waters of the Beuvronne and the Ourcq to bring them to the village of La Villette. From there, a distribution basin could feed the public fountains while two canals would allow the joining up of the Seine to the Seine, between the Arsenal and Saint-Denis.

The Revolution delayed the works which were then reconsidered in 1802. The engineer Girard was entrusted with them by the Administration of Napoleon. Work only began in 1806. This gigantic enterprise, the basin of the Villette measures seven hundred metres by seventy, was led with the help of one thousand five hundred Russian and Austrian prisoners of war. The basin of the Villette was completed in 1808 and inaugurated by the Emperor.

The navigation of barges began on the Canal of the Ourcq in 1813.

The Canal Saint-Denis between La Villette and the Seine at Saint-Denis, 7.5 kilometres long and with seven locks, was started in 1811 and completed during the Restoration.

In 1815, construction of the Canal Saint-Martin was also considered.

The canal was opened to river traffic on 14th November 1825.

The Bassin de l'Arsenal, circa 1880.
A painting by Pierre Vauthier.

With a length of 4,553 metres, the canal had a height difference of 25.25 metres regulated by eight locks (two were removed in 1861). At the beginning, five stone bridges, two foot-bridges and seven swing bridges crossed the canal.

The works on the Boulevard du Prince Eugène (Voltaire) and de Ménilmontant (Avenue de la République) led Haussmann to buy back the concession of the Canal which became the property of the Town in 1861. At that time, it was in the open air. The creation of new, wide avenues, completed in 1867, resulted in the partial covering of the work. A wonderful basket-handle arch,

1,800 metres long, was built between the Bastille and the avenue de la République.

At the same time that the canal de l'Ourcq and the Bassins de la Villette were being dug, Napoleon I had the Arsenal basin fitted out and deepened to be used as an outer port for the boats which would pass on the Canal Saint-Martin. It is the only remaining port in Paris, whose surface and general configuration has not changed since the time of its creation. 650 metres long and 60 metres wide, it replaced, in 1806, the ancient moat skirting Charles V's ramparts.

For a century and a half upto the 1980s, the very wide port, situated on the east bank, below the Boulevard de la Bastille, was permanently in use.

The basin of the Arsenal spanned by an iron foot-bridge became a pleasure boat harbour on the initiative of Jacques Chirac, the Mayor of Paris.

The lock of the Pont Morland in 1885.
A water-colour by Fraipont.

The Bassin de l'Arsenal in 1993.

A riverboat «Paris-Canal» in front of the Hôtel du Nord.

A garden with a walk has been laid out on the low wide bank facing south-west.

For more than a century, the Canal Saint-Martin played a considerable rôle in the port traffic of Paris. It was also, to the east of the town, a colourful space with light and an atmosphere that delighted painters, novelists, poets and film-makers. In a very famous film by Marcel Carné «Hôtel du Nord», Louis Jouvet was answered scathingly by Arletty... about the «atmosphère».

In the 1960s, once the river traffic had considerably decreased, engineers and technocrats suggested replacing this canal by an expressway called the «North-South Axis»... The relentless opposition by associations for the protection of this site caused the abandonning of this project after a dozen years.

Since that time, the inscription in the Inventaire des Sites has guaranteed the protection the «Grand Canal de Paris» and its surroundings. The banks have been restored, the Bassins de l'Arsenal and de la Villette have been transformed into marinas and tree-lined walks.

In 1993, the «Direction des Parcs et Jardins de la Ville de Paris» undertook the transformation of the central reservation of the Boulevards Richard Lenoir and Jules Ferry. A new, 1,800-metre long garden is being laid out between the Place de la Bastille and the Rue du Faubourg du Temple, by the architects David Mangin and Jacqueline Osty.

The Canal Saint-Martin begins at the far end of the Bassin de l'Arsenal on the Place de la Bastille side. It immediately disappears into a long vaulted passage, nearly 2 kilometres long; one of the most unusual and magical places in Paris.

The long Richard Lenoir-Jules Ferry vault ends at the Rue du Temple where the canal reappears.

A stone-sawyer on the edge of the basin, circa 1880. A drawing by Fraipont.

The succession of locks which control the passage between the different levels from the Rue du Faubourg du Temple to La Villette, offers a most surprising and pleasant town landscape.

A double lock separates the tunnel from the «Bassin de la Douane»; two elegant iron foot-bridges span this basin, while a swing bridge joins the foot-bridge of the Rue Dieu. The unpleasant-looking Richerand foot-bridge, the third on the way upstream, rebuilt in the sixties, should be replaced.

Opposite the Rue de la Grange aux Belles, there is a second swing bridge for cars and also the fourth foot-bridge called «la Grange aux Belles» at right angles with which begins the double lock with the same name.

In the area delimited by this street, the Canal Saint-Martin and the Rue des Ecluses Saint-Martin, there was, in the past, the famous Montfaucon Gallows where the condemned were han-

A foot-bridge facing the Hôtel du Nord.

A river-boat of the Canauxrama Company.

The «Canotier» at the first lock of the Villette.

ged and remained displayed. Erected in the XIIIth century, the gallows of Paris were dismantled in 1760.

The Ville de Paris ought to give the names of Arletty, Marcel Carné and Louis Jouvet to these foot-bridges and locks since you are, here, just in front of the famous «Hôtel du Nord» at n°112 Quai de Jemmapes, recently restored and included in the Inventaire des Monuments Historiques.

The Jemmapes foot-bridge links the Rue des Récollets on the left and the Rue Bichat on the right, above the lock, in the elbow of the canal which widens in the direction of the large Bassin des Récollets.

A double lock separates the Bassin des Récollets from the Bassin du Combat, at right angles with the bridge linking the Rue des Ecluses Saint-Martin to the Rue Eugène Varlin.

The Bassin du Combat gets its name from the presence in the past of the barrier of Combat (Place du Colonel Fabien). At the end of the XVIIIth century, animal fights (fights between dogs and wolves, dogs and bears, dogs and wild boars and also fights between dogs and bulls) were organized here. After the Reign of Terror, they came into fashion and remained until 1850.

Upstream the canal disappears under the Boulevard de la Villette to reappear at the corner of the Quai de la Loire and the Avenue Jean-Jaurès where a double lock links it with the Grand Bassin de la Villette, whose surface area is three times that of the Palais Royal.

Very important works were carried out in 1987-1988 on the site of the Place Stalingrad. Supervised by the architect, Bernard Huet, they enhanced the central element on which all the town planning of the quarter is oriented: the Rotonde de la Villette built by Claude-Nicolas Ledoux in 1787. The Direction de l'Eclairage du service de la Voierie de la Mairie de Paris recently installed flood-lighting on this monument.

At the time of the construction of the Wall of the Farmers General, this rotunda was the biggest of three buildings known as «Saint-Martin gates». It was accompanied by two other buildings about a hundred metres apart.

At the beginning of the «Route d'Allemagne» there was the «Pantin» barrier. At the beginning of the «Route des Flandres» there was the «la Villette» barrier. They remained in place until 1871.

The Bassin and the Rotonde de la Villette.

The Rotonde de La Villette

The Rotonde is a large building on two storeys. The quadrangular ground floor is preceeded on its four sides by a monumental projecting porch formed by eight square section pillars; they bear the entablature of a very flat triangular pediment. The upper storey is the rotunda itself, characteristic with its circular gallery of forty Doric coupled columns bearing the semicircular arches, themselves topped by an attic with twenty square skylights. The building is crowned with a beautiful Doric cornice.

The City of Science and Industry: open daily 10am to 6pm. Closed on Mondays; Entrance: 30 Avenue Corentin-Cariou; Tel: 40 05 80 00. Maison de la Villette, parc de la Villette. Information centre and activities. Open every day except on Mondays from 1pm to 6pm. Tel: 40 03 75 10.

The Rotonde de la Villette remains, with the barriers of the Trône, Enfert and Montceau one of the rare relics of the forty-seven monumental gates built around Paris by Claude-Nicolas Ledoux.

The Bassin de la Villette was, at the beginning of the XIXth century, a place of entertainment. In February 1827, for forty sous an hour, ladies could be driven there, sitting on sledges. In Summer, water-tournaments were organized. The bassin became fashionable on the day of its inauguration, on 2nd December 1808, on the anniversary of the Coronation of the Emperor.

Beyond the 70-metre wide channel that connects the large Bassin de la Villette with the narrower 30-metre wide one which leads to the junction of the canals, there is the lifting bridge of the Rue de Crimée and the adjoining foot-bridge.

Machinery of the lifting bridge of the Rue de Crimée.

This work, unique in Paris, was built at the end of the XIXth century. The deck rises to let boats pass.

Beyond the bridge, on the left side, facing the Place du Bitche, Saint-Jean - Saint-Christophe de la Villette Church was built in 1844 by Eugène Lequeux. There is a public garden planted with catalpas and also a nice bandstand in front of the façade. After this you arrive at the junction of all the canals; a truly magnificent space.

On the left, the entrance of the Canal Saint-Denis is controlled by a double lock in the centre of which stands an unusual and picturesque lock-keeper's house. Built recently on the left bank of the Canal de l'Ourcq by the architect Adrien Fainsiller, are the huge buildings of the Museum of Science and Industry, of which the Géode has become the symbol. On the opposite bank, on the right side, there is the park designed by the landscape gardener, Bernard Tschumi.

There are only a few elements left of the gigantic old buildings of the cattle market and slaughter-houses of la Villette built by the architect Janvier, on the initiative of the Prefet Haussmann and Emperor Napoléon III. There are the Veterinarians' Rotunda near the Science and Industry Museum and the Great Hall in the park. 242 metres long and 86 metres wide, composed of seven naves delimited by a forest of cast iron columns supporting the roofs cut with long glass tympana. This magnificent Hall is today used for exhibitions and shows.

The City of Science and the Géode de La Villette. Architect : Adrien Fainsilbert.

There are boat-trips between the Port de l'Arsenal and the Bassin de La Villette all year round on board the glass-roofed boats of the Canauxrama Company, one of them naturally called Arletty after the great comedian.

There are departures every day at 9am, 2.45 and 3pm from the Bassin de la Villette - 5 bis Quai de la Loire, one hundred metres from the Jean Jaurès Metro Station - and in the opposite direction at 9.45 a.m. and at 2.30 p.m. from the Port de l'Arsenal - on the Boulevard de la Bastille side, near the Bastille Metro Station. Tel: 42 39 15 00.

Canauxrama Company also organizes trips on the Canal de l'Ourcq the from the Bassin de La Villette to Meaux (Vignely lock). These trips are everyday, except on Wednesdays, from March 15th to November 15th, by reservation.

Short trips are organized from March 15th to November 15th, alternating between the Bassin and the Parc de La Villette, on weekends and during school holidays, every half hour between the landing stage at 5 bis Quai de la Loire and the information centre «Villette» towards the park.

The «Paris-Canal» Company, 21 Quai de la Loire, Tel: 42 40 96 97 offers three-hour trips between the Orsay Museum and La Villette Park along the Saint Martin Canal. Embarkation 9.30am, 2.25pm at the Quai Anatole France; 10am and 2.30pm at La Villette Park.

«Paris-Canal» offers one-hour trips to discover the Bassins de La Villette, the Park and the City of Science. This company also offers trips between the Quai Anatole France and the loop of the Marne, every Sunday in June, July, August and September, departure: 9.30am; return: about 5.30pm. Trips can be organized on weekdays and by request for groups.

Bibliography

La Seine à travers Paris, Jean Saint-Juir et Georges Fraipont, 1890.
Atlas des voies navigables de la France, Imprimerie Nationale, 1896.
Le port de Paris, Librairie Félix Alcan, 1911.
Paris sur Seine. Féerie des vingt arrondissements, Alexandre Arnoux, Grasset, 1939.
Paris, ma grande ville, Alexandre Arnoux, Grasset, 1939.
Fenêtres sur Seine, Jean Prasteau, Editions du Palais Royal, 1961.
Les ponts de Paris, Renée Plouin, Editions Olivier Perrin, 1967.
Ponts de Paris à travers les siècles, Henri-Louis Dubly, Editions H. Veyrier, 1973.
Le canal Saint-Martin et la Villette, Rapport, Marc Gaillard, 1977.
Quais et ponts de Paris, Marc Gaillard, Editions du Moniteur, 1981.
Paris sur Seine, François Beaudoin, Editions Nathan, 1988.
Histoire des transports parisiens, Marc Gaillard, Editions Martelle, 1990.

The documents illustrating this book belong to the archives of the publisher Joël Bongini and the author. Photographs by Alberic Guernon and from the Bulloz photo library.

Acknowledgments :

We arc grateful to the following for permission to reproduce textual and photographic material: les galeries Charles et André Boully, Alain Letailleur, Thierry Salvador, Robert Schmit, Roger et Mancheron, de Rohan, Berko, Odermatt, Brame et Lorenceau à Paris : la galerie Richard Green in London, Serge Belloni,

Directeur d'Edition : Joël BONGINI
ISBN 2-87890-036-7

BP 0540, 3, rue des Vergeaux
80005 AMIENS Cedex

Conception : J. Bongini
Translate from French into English by
Liliane and Ronald Brown and Richard Driscoll.

Réalisation - Montage - Photogravure noir et blanc :

trait & compo

77 74 58 99

Photogravure couleur :
BDP Scann
77 36 22 22

Impression :
Saint-Paul France S.A., Versailles
Dépôt légal : juin 1994
N°6-94-0811